Population and Community Biology

COMPETITION

Population and Community Biology Series

Principal Editor

M. B. Usher
Reader, Department of Biology, University of York, UK

Editors

M. L. Rosenzweig
Professor, Department of Ecology and Evolutionary Biology, University of Arizona, USA

R. L. Kitching
Professor, Department of Ecosystem Management, University of New England, Australia

The study of both populations and communities is central to the science of ecology. This series of books will explore many facets of population biology and the processes that determine the structure and dynamics of communities. Although individual authors and editors are given freedom to develop their subjects in their own way, scientific rigour is essential and often a quantitative approach to analysing population and community phenomena will be used.

Population Dynamics of Infectious Diseases: Theory and applications
Edited by R. M. Anderson
Hardback (0 412 21610 8), 368 pages.

Multivariate Analysis of Ecological Communities
P. Digby and R. Kempton
Hardback (0 412 24640 6) and paperback (0 412 24650 3), 206 pages.

The Statistics of Natural Selection
Bryan F. J. Manly
Hardback (0 412 25630 4) and paperback (0 412 30700 6), 484 pages.

Food Webs
Stuart L. Pimm
Hardback (0 412 23100 X) and paperback (0 412 23110 7), 220 pages.

Predation
Robert J. Taylor
Hardback (0 412 25060 8) and paperback (0 412 26120 0), 166 pages.

COMPETITION

Paul A. Keddy

Department of Biology
University of Ottawa

London   New York
CHAPMAN AND HALL

First published in 1989 by
Chapman and Hall
11 New Fetter Lane, London EC4P 4EE
Published in the USA by
Routledge, Chapman and Hall
29 West 35th Street, New York NY 10001

© *1989 Paul A. Keddy*

Typeset in 10/12 Times by
Thomson Press (India) Ltd, New Delhi
Printed in Great Britain by
St. Edmundsbury Press Ltd
Bury St. Edmunds, Suffolk

ISBN 0 412 31350 2 (hb)
ISBN 0 412 31360 X (pb)

British Library Cataloguing in Publication Data

Keddy, Paul A., 1953–
  Competition.—(Population and community
  biology).
  1. Competition and cooperation
  I. Title II. Series
  302.1'4

  ISBN 0 412 31350 2
  ISBN 0 412 31360 X pbk

Library of Congress Cataloging in Publication Data

Keddy, Paul A., 1953–
  Competition/Paul A. Keddy.
    p. cm.—(Population and community biology series)
  Bibliography: p.
  Includes index.
  ISBN 0 412 31350 2.—ISBN 0 412 31360 X (pbk.)
  1. Competition (Biology)   I. Title. II. Series.
  QH546.3.K43 1989
  574.5'247—dc19   88-34309
                                                        CIP

Contents

My view, based on long and painful observation, is that professors are somewhat worse than other people, and that scientists are somewhat worse than other professors. Let me demonstrate that these propositions are self-evidently true. The foundation of morality in our society is a desire to protect one's reputation. A professor's reputation depends entirely upon his books and his articles in learned journals. The narrower the field in which a man must tell the truth, the wider is the area in which he is free to lie.

R. M. Hutchins (1963)

Preface

This book has two principal objectives. The first is to provide an overview of existing knowledge about competition. The second is to organize this knowledge in such a way that new research paths are suggested. Such a treatment of competition is badly needed. Although there is a voluminous literature on the topic there is no recent synthesis to which experienced researchers or new students may turn. This is my attempt to provide such an overview. I have tried not only to summarize what is known, but also to stress the unknowns in the hope that some new and innovative research will result. A book such as this faces two challenges at the outset: the sheer volume of the literature, and the presence of established research traditions which determine how that literature is to be interpreted and understood.

The literature on competition is as vast and diverse as beetles in the biosphere. How better to begin, then, than with the preface from Crowson's (1981) volume on the Coleoptera? He observed:

> To deal with so vast a group as the Coleoptera...is doubtless an over-ambitious aim for any single author; it is inevitable that my attempt to do so will not satisfy specialists in their own particular fields. I hope, however, that such specialists, once they have overcome their initial dissatisfaction, may gain from this book by coming to see their particular interests in wider contexts, and perhaps even by picking up ideas which might suggest new and fruitful directions for their investigations.

In addition to its vastness, the literature on competition is strongly coloured by tradition. For example, communities structured by niche differentiation have received far more attention than those structured by dominance. Intraspecific competition in monocultures has received more attention than diffuse competition in multispecies communities. Studies of birds are far more numerous than of fungi, particularly when the birds occur on tropical islands. There are many special cases and few general principles. The material in this book reflects my attempts to counterbalance some of these traditions. For example, dominance-structured communities, multispecies experiments and empirical studies of general principles all receive increased emphasis.

In an area like the study of competition, with such a vast literature and well-established research traditions, a book which simply repeated what others have said and described the status quo would not only be of limited value, it would also probably be boring. This is my attempt to review what needs to be reviewed, to highlight areas which have received inadequate consideration, to describe unresolved problems and to suggest avenues which might eventually allow someone to write a book entitled *Competition Theory*.

This book therefore does not emphasize evolutionary ecology. Evolutionary ecology provides only one of several possible perspectives upon competition, yet an overwhelming majority of studies use this conceptual framework; at least three recent books (Roughgarden, 1979; Pianka, 1983; Arthur, 1987) and a review (Arthur, 1982) have summarized this view of the discipline. At the very least one could argue that the time has arrived to explore other complementary contexts for research. Moreover, although Darwin's concept of evolution through natural selection revolutionized biology a century ago, it is now frequently used in uninspired ways to explain how yet another set of observations are adaptive. Instead of offering major new insights, such studies simply provide new data which are interpreted in light of the existing paradigm (*sensu* Kuhn, 1970). Yes, populations of organisms evolve. Yes, many interactions are shaped by evolution. However, beyond elaborating the details of this process, what general, operational, predictive statements can we make about the way in which competition structures animal and plant communities? What approaches have offered new insights to these questions? What research avenues have been overlooked? Why has there been so little apparent progress since the pioneering work of Clements and Gause? These are the questions that directed this book.

As the proximate cause of the book, I must be held responsible for errors, omissions and oversights. A book, is, however, a product of circumstance, and who better than an ecological audience to appreciate the complexity of interactions which can create a particular situation? Canadian authors are said to be strongly influenced by the land and the isolation of small-town life; undoubtedly my style has been influenced by early schooling on the prairies, where intellectual inclinations and lice had the same popularity. My early interests in science and natural history were strongly encouraged and supported by my parents, Norm and Jean Keddy. Bruce McBride not only encouraged my interests in herpetology, but also suggested I apply to work as a park naturalist. Dan Strickland, Chief Naturalist in Algonquin Provincial Park, hired me, and he and the other naturalists there can probably be detected somewhere in this text too. Duncan Cameron and Michael Boyer, at York University in Toronto, provided the encouragement and the academic counterpoint to my training as a naturalist. My PhD supervisor, Chris Pielou, permitted individualism to flourish. My colleague Doug Larson, at the University of Guelph, was willing to debate nearly anything and offer well-considered alternatives to my ideas. Phil Grime reminded me that the important questions are the big ones. Scott Wilson is undoubtedly in here too, because of the many hours we have spent discussing these issues around campfires, at conferences and in bars. Finally, Michael Usher, at the University of York, is the next-to-proximate cause because he first suggested that I write this book, and provided advice and encouragement as it progressed.

Each of the chapters ends with questions for discussion. I hope that these will generate the sort of lively debate that we had in graduate courses and in seminars at the University of Ottawa. Here I must acknowledge my recent graduate students, Scott Wilson, Bill Shipley, Connie Gaudet, Dwayne Moore and Irene Wisheu, as well as my colleague David Currie. One cannot work with people for years without absorbing some of their ideas and opinions. Where I have unconsciously borrowed from their minds, I only hope I have offered something worthwhile in exchange. I also thank Anita Payne for her help over the past year, particularly with the literature searches for the analyses in Chapter 8. I also thank Jacques Heli for preparing the figures. James Brown, David Currie, Jared Diamond, Gray Merriam, Bill Shipley and Richard Southwood graciously and promptly commented upon individual chapters. I particularly thank Steve McCanny, Rob Peters, Scott Wilson and Michael Usher for their heroic work in reading an early draft and providing insightful, humorous and constructive criticisms. Perhaps I should have incorporated more of them.

Introduction

The living system of which we humans are a part has apparently unending cycles of birth, growth, consumption, reproduction, defecation, death and decay. In the midst of this at least one species, *Homo sapiens*, attempts to find patterns that organize perceptions which otherwise would be chaotic. This tendency to organize and find pattern appears to be a fundamental trait of ourselves, both as individuals and as a species.

The concept of self and other seems to emerge as one of our earliest classification schemes. We are told that infants do not even perceive that they are different from their parents, but by 2 years of age 'selfhood' is a part of human mental structure. As the idea of self matures, we see in nature many other units which persist through time, seem to have some sense of self, eat, defecate and reproduce. So it is natural to begin by describing our living surroundings as being made up of many individual organisms. We should recognize that this in itself is arbitrary, but it seems to work, so we continue to do it.

Once we break the living systems around us into 'individuals', the whole structure of modern biology rapidly emerges. Since some individuals are more similar than others, taxonomy develops. Since individuals can be pulled to pieces, anatomy, morphology and physiology become established. Since individuals appear to encounter each other and interact, with different consequences, ideas such as sex, predation, mutualism and competition emerge. It is this last great interaction among individuals which we explore in this book.

1 Studying competition

There is some danger that a symposium on competition
which begins with a section on definitions may so
irritate later speakers that the whole meeting
degenerates into a display of semantics.

J. L. Harper (1961)

Clearly the need is not only for a strict definition
of competition but also for a discerning interpretation
of the definition.

A. Milne (1961)

We start with a word whose meaning we think we
understand... and begin to investigate the things
which it designates. We always find that it changes
its meaning in the course of the investigation, and
sometimes we have to invent new words for the things we
discover.

J. B. S. Haldane (1985)

How every fool can play upon the word!

W. Shakespeare, The Merchant of Venice

Why study competition? As one of the three major kinds of biotic interactions
(along with predation and mutualism), we may anticipate that it could be a
force as fundamental to ecosystems as gravity is to planetary systems. At this
stage the study of ecosystems is not unlike the pre-Copernican study of the
Solar System. We have a wealth of detailed observations on the natural history
of our planet, but are only beginning to uncover (or invent) the general
principles which can organize this mass of observations. One of the features
that makes ecology and the study of competition so exciting is the period of
rapid development we are in. The founding fathers of physical theory, such as
Copernicus, Galileo, Newton and Einstein, are all dead. The founding fathers
of ecology may be with us now as graduate students or professors.

There is presently no coherent body of 'competition theory', although there are concepts and models which invoke the existence of competition as a shaman invokes a local spirit. There is, however, a rich mixture of observation, experimentation, speculation, concept, theory and models with which ecologists can work. If these can be organized in a coherent manner, then we may have at least the foundation of such a theory.

The value of a concept like competition can be judged only by its contribution to the development of ecological theory. How can such value be assessed? First, if the concept is useful, it should allow our minds to organize and understand apparently chaotic displays of nature as perceived through our limited sensory apparatus. If we can organize some of this apparent chaos with the concept, then we can be said to in some way understand it, in that we can carry about an intellectual framework which appears consistent with patterns that we observe in nature. If this organization succeeds, then we should be able to predict – that is, given existing states and our knowledge, we should be able to foresee our world accurately. This is not only a goal sought by scientists: it seems to be basic to human nature, since diviners, oracles, prophets, mystics and necromancers can be found throughout human history. Roget's Thesaurus (Chapman, 1977) lists 87 forms of divination, from aeromancy to zoomancy.

To begin studying nature we must attach names to objects and phenomena. This provides the vocabulary for exchange of ideas. As the Bible observes (John 1:1): 'In the beginning was the Word'. Although we must begin by defining the meanings of our words, words are only human concepts imposed upon nature. When they help us to communicate, it is excellent; when they become barriers between ourselves and real phenomena, or lead to endless debates of terminology, they can be replaced or simply discarded. Some ambiguity will always remain with words, and it is to be expected that words and their meanings will evolve along with the discipline itself, as in the introductory quotation from Haldane.

1.1 A DEFINITION OF COMPETITION

Definitions of competition present a challenge. It may be difficult to find a definition that is both sufficiently precise to satisfy the most meticulous personality, and sufficiently robust to encompass the riotous display of possibilities in nature. The definition may emphasize the postulated mechanism of the interaction, or it may be more operational and emphasize responses to experimental perturbations. A perusal of recent textbooks will reveal a wide array of attempts to satisfy these conflicting objectives. Some authors no longer even use the term. For the purposes of this book, **competition** will be defined as

the negative effects which one organism has upon another by consuming, or controlling access to, a resource that is limited in availability.

This definition of competition provides a starting point for exploring nature. This chapter begins with a brief historical account, and then explores the many different kinds of competition which could exist in nature. This requires explicit consideration of: (1) kinds of resources; (2) mechanisms of competition; and (3) the kind of individuals of groups competing.

1.2   OTHER VIEWS ON THE DEFINITION OF COMPETITION

Harper (1961) and Milne (1961) have reviewed the use of the word competition by botanists and zoologists. Harper observes that 'competition' is the response of plants to density-induced shortages, and proceeds to consider definitions used by agronomists, ecologists and geneticists. The agronomist is primarily concerned with the way in which a crop exploits the resources in an environment. Usually this work follows two steps: a description of density-dependent effects and an analysis of causes. This is frequently accomplished by exploring the way in which crop yield varies with sowing density, often with a range of fertilization levels. Harper notes that the population ecologist has less control over the system being studied, and frequently is interested in processes occurring over longer time-scales, than is the crop scientist. He observes that ecologists are concerned with 'those hardships which are caused by the proximity of neighbors', and suggests that 'interference' is a preferable word.

Milne (1961) reviews the historical contortions and confusion which have surrounded the use of the word competition, and concludes that we have three courses: accept ambiguous use, drop it altogether or provide a restrictive definition: 'Competition is the endeavour of two (or more) animals to gain the same particular thing, or to gain the measure each wants from the supply of a thing when that supply is not sufficient for both (or all)'. He concludes with the appeal: 'Clearly the need is not only for strict definition of competition but also for a discerning interpretation of the definition'.

In order to test hypotheses about competition, it is important to emphasize the operational parts of the definition. Recall (p. 2) that competition is an interaction in which individuals (genotypes, populations): (1) have negative effects upon each other (2) by influencing access to resources. The first part, assessing negative effects upon each other, seems straightforward for measurement and testing, and some examples of this are given in Chapter 2. Whether both must be negatively affected is open to question, however, since in very asymmetric interactions the dominant may be so little affected by the subordinate that negative effects upon it cannot be detected; for convenience, we may regard such circumstances as a limiting case. The second part of the definition, demonstrating the mechanism causing the negative effects, is more difficult. Many animal ecologists insist that resource limitation must be demonstrated before one can conclude that competition is occurring (for example, Milne, 1961; Price, 1984a), but others have argued for a more operational approach (Wilbur, 1972). Plant ecologists in general have remained more operational (Fowler, 1981; Silander and Antonovics, 1982; S. D. Wilson and Keddy, 1986a), but Tilman (1987a) presents an opposing view: the debate over the meaning of competition in plant communities is on-going (Thompson, 1987; Tilman, 1987b; Thompson and Grime, 1988).

Debates over definitions themselves may accomplish little, and even their entertainment value is limited, since the controversies ignited may detract from important questions rather than stimulate them. All human concepts are only limited attempts to organize a complexity beyond the organizational capacities of our nervous system, so we should be realistic about why we need definitions, and proceed with the task at hand – to use the definition as an initial reference point for studying nature. We can expect our definitions to evolve as we learn more about the phenomenon itself (Haldane, 1985).

1.3 KINDS OF RESOURCES

A good general definition of a **resource** might be: 'any substance or factor which is consumed by an organism and which can lead to increased growth rates as its availability in the environment is increased' (Tilman, 1982). It is not necessary to be more specific at this point; it could be growth rates of individuals, genotypes, populations or species, depending upon the scale of the investigation. There are only a few basic elements serving as resources for living organisms (Table 1.1), and for this reason Morowitz (1968) has

Table 1.1 Major elements required by living organisms and their functions

Element	Function
C	Structure; energy storage in lipids and carbohydrates
H	Structure; energy storage in lipids and carbohydrates
N	Structure of proteins and nucleic acids
O	Structure; aerobic respiration for energy release
P	Structure of nucleic acids and skeletons; energy transfer within cells
S	Structure of proteins

described this planet as one made up of CHNOPS life forms. These major elements which comprise living organisms are the most obvious place to begin exploring 'resources' which individuals, genotypes and populations accumulate, and for which they may potentially compete.

Price (1984a) proposes that any study of a population or community should begin by explicitly considering resources. He argues that resources will always play a role in organizing communities, whereas other factors like weather or predation, although important in some cases, are not inevitably significant.

Moreover, if one accepts that resource limitation must be proven to demonstrate competition, then his admonition is particularly relevant to studies of competition. It is therefore important to explore kinds of resources further. Four schemes follow: (1) trophic position; (2) temporal and spatial distribution; (3) mode of consumption; and (4) resource ratios.

1.3.1 Trophic position

Although the elements comprising living organisms (Table 1.1) may be considered the most fundamental resources, they may be consumed in very different ways by different organisms. It is sometimes useful to break down potential consumers into two groups: the autotrophs and the heterotrophs. Autotrophs assimilate raw materials which, in the presence of sunlight, are converted to more-complex molecules with higher potential energy. In contrast, heterotrophs assimilate other organisms rather than raw materials. The chemical composition of different consumers is remarkably constant (Table 1.2), however, so the distinction between autotroph and heterotroph is simply the size, shape, composition and distribution of the packages in which the resources are acquired.

Table 1.2   Atomic composition of four typical CHNOPS organisms (after Morowitz, 1968, Table 3.2)

Element	Mammal[a]	Vascular plant[b]	Arthropod[c]	Moneran[d]
C	19.37	11.34	6.10	12.14
H	9.31	8.72	10.21	9.94
N	5.14	0.83	1.50	3.04
O	62.81	77.90	79.99	73.68
P	0.63	0.71	0.13	0.60
S	0.64	0.10	0.14	0.32
Total	97.90	99.60	98.16	99.72

[a] Humans; [b] alfalfa; [c] copepod; [d] bacteria.

1.3.2 Temporal and spatial distribution

Price (1984a) classifies resources by the way in which they vary with time, and recognizes five kinds. Four of these are shown in Fig. 1.1. Increasing resources are those which gradually increase over the active season of an organism and then suddenly decline. Decreasing resources are produced suddenly at the beginning of a season, and then gradually decline. Pulsing or ephemeral resources increase rapidly and then decline rapidly. Steadily renewed resources are continuously renewed over long periods. Constant resources, not shown in Fig. 1.1, are physical in nature, and are largely unaffected by seasonal change. Price uses this last category to deal with space as a resource for sessile organisms; since disturbance and death continually create new spaces for colonists, space can be treated as another example of a continually renewed resource. Price's five resource types can therefore be reduced to four. Figure 1.1 illustrates this modified classification, and Table 1.3 gives examples of organisms using these resources.

If uniform and patchy distributions of each kind of resource supply type are recognized, then it doubles the number of kinds of resources in Price's scheme. Southwood (1977) produced a more elaborate classification which included both spatial and temporal variation. He recognized four temporal categories and three spatial categories. Combining the two would produce 12 different

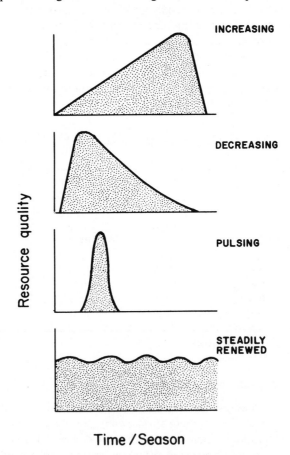

Figure 1.1 Four kinds of resources classified by pattern of temporal variation (modified from five types presented by Price, 1984a). Examples of each type are given in Table 1.3. This scheme does not consider spatial variation or chronically low resource availability.

Table 1.3 Four types of resources, examples of them, and examples of organisms that exploit them (modified and expanded from Price, 1984a)

Resource type	Resource	Exploiter
Increasing	Temperate forest foliage	Herbivores
	Flowers	Generalist pollinators
	Temperate-zone insects	Insectivores
Decreasing	Seeds in deserts	Granivores
	Insect stages	Specialized parasitoids
	Soil moisture	Temperate forests
	Temporary ponds	Amphibians
	Litter in deciduous forests	Decomposers
Pulsing	Phytoplankton	Zooplankton
	Herbaceous plant parts	Specialist herbivores
	Flowers in temperate regions	Specialist pollinators
Steadily renewed	Gut contents, blood	Internal parasites
	Marine plankton	Intertidal filter feeders
	Flowers in wet tropics	Pollinators
	Foliage in wet tropics	Herbivores
	Space	Intertidal organisms
	Space	Plants
	Nesting sites	Hole nesters

resource types. Southwood's classification was actually designed to explore the evolution of life-history traits, and was concerned with many attributes of the favourableness of habitats beyond solely resource supply; if, however, we take his measure of habitat favourableness and transform it to a measure of resource availability, then his classification can be applied to recognize different resource types (Fig. 1.2).

A problem with both of the above classification schemes is that they do not consider the distinction between chronically high and low supply rates. It is arguable that resources with high renewal rates have consequences fundamentally different from those with low renewal rates (for example, Grime 1977, 1979; Greenslade, 1983; Southwood, 1988). The classification of resources in Fig. 1.2 therefore includes a category for constant but chronically low resource supply rates. Spatial and temporal scales are, of course, only meaningful when measured relative to the scale of the organism concerned. For simplicity, scaling terms like generation time and foraging range have been left off Fig. 1.2; more-complete accounts of scaling are given in Southwood (1977, 1988) and Begon and Mortimer (1981).

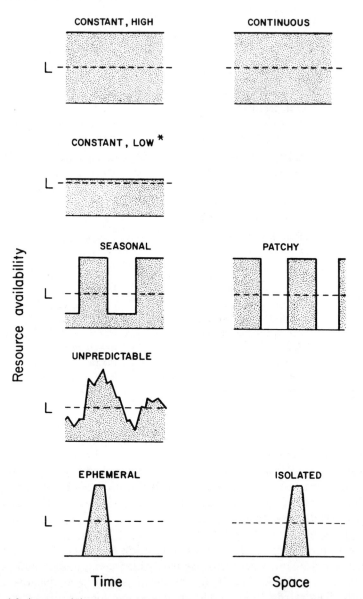

Figure 1.2 A more elaborate classification of resource types, based upon variation in resource availability in time and space. Five temporal distributions can be combined with three spatial distributions for a total of 15 possible resource types. L indicates the level of resource necessary for reproduction. This is modified from Southwood (1977) by the addition of the 'constant low' category mark with an asterisk. For simplicity, scaling terms like generation time and foraging range have been left off the figure, but clearly the axes are only relevant on scales adjusted to the life-history characteristics of the species concerned.

1.3.3 Mode of consumption

Yodzis (1986) proposes two fundamentally different types of resource consumption. Some organisms harvest a fraction of a resource over a large area. This leads to consumptive competition. Others harvest all a resource from a fraction of the area. This is space competition. Blue whales and Sequoias might illustrate these foraging strategies, in that a Blue whale removes a fraction of the available krill from a large area of ocean, whereas a Sequoia removes a large fraction of the available light from a small area of land. However, the validity of this classification may well vary with type of resource, for Seqouias would be more like whales if their growth were limited by carbon dioxide levels in the atmosphere as opposed to light.

These different modes of consumption would lead to distinct types of community organization. Consumptive competition would produce communities organized by resource partitioning, whereas space competition would produce communities organized by dominance hierarchies.

1.3.4 Resource ratios

As Tables 1.1 and 1.2 suggest, organisms must frequently forage for more than one resource. Tilman (1982), building on MacArthur (1972), emphasizes the responses of organisms to ratios of resources. He proposes a distinction between substitutable and essential resources. These concepts are best understood by considering the situation where an organism is foraging for two different resources. Two resources are considered substitutable if a population can maintain its growth rate by substituting one resource for the other at all abundances of the two resources. In contrast, essential resources cannot be substituted for each other, and the growth rate of a species will be determined by either one or the other resource, whichever is less available.

This distinction is illustrated in Fig. 1.3. A species' response to the two resource axes is shown with a growth isocline. At all points along this isocline (i.e. at all combinations of resources 1 and 2 along the line) the population growth rate, dN/dt, is constant. It may be easiest to consider the case where the line represents the zero net growth isocline. In this case all resource combinations in the shaded areas result in declining population growth.

A further possible distinction is the division of substitutable resources into three categories: perfectly substitutable, complementary and antagonistic. **Perfectly substitutable resources** produce straight isoclines which intersect both axes. Predators feeding on closely related groups of organisms would illustrate consumption of perfectly substitutable resources. For example, predatory mammals might be expected to treat congeneric *Ambystoma* salamanders as substitutable, and owls may find congeneric *Peromyscus* mice just as substitutable. If resources are perfectly substitutable, then it may make little sense to treat them as different resources. Which conceptual framework – that of the predator, or that of the taxonomically skilled

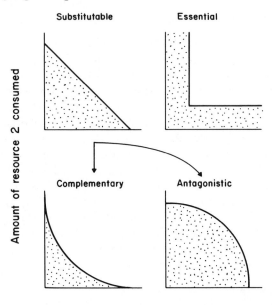

Figure 1.3 A classification of resources based upon effects of consuming different ratios of resources. The solid line gives the amounts of the two resources which produce zero growth (zero net growth isoclines) and the shaded areas give regions where growth would be negative. Since all organisms require at least two resources, each pair of resources consumed can be placed into one of these categories. (After Tilman, 1982.)

ecologist – is really more useful? **Complementary resources** have a synergistic effect, such that a mixture of the two enhances growth more than equivalent amounts of either resource by itself. A good example of this comes from human vegetarian diets. Amino acids necessary for building proteins exist in different ratios in different foods. By combining foods such as rice and beans, the usable protein content can be increased as much as 40% over what would be usable had the foods been eaten independently (Lappe, 1971). **Antagonistic resources** interfere with one another when consumed, so that a mixture of the two reduces growth relative to a diet of either resource by itself. Such a situation might occur if each resource required different enzymes or gut morphologies for digestion and assimilation, thereby preventing an organism from efficiently consuming either in mixtures. The many examples of chemical defense in animals (for example, Blum 1981) and plants (for example, Wallace and Mansell, 1976; Coley *et al.*, 1985) present opportunities for such antagonistic interactions. In such cases a diet specialized on one resource or the other would seem to be beneficial.

For animals, which consume resource packages, resource axes could be labelled as kinds of prey or different components of them. Two similar kinds of prey might be considered substitutable packages, providing one conceptual framework for studying the foraging of the predator. However, if vitamins in the prey are considered separately, then the predator might be modelled as foraging for essential resources.

1.3.5 Consequences of such classifications for competition

Based on this introductory view of CHNOPS resources, one could already make a prediction about the way in which competition could evolve among different kinds of organisms. Because they acquire resources in packages which can differ radically in size, shape, composition and distribution, animals may be expected to specialize to consume a comparatively small subset of resource packages. In contrast, plants seem likely to share a requirement for basic resources, with limited options for specializing upon a subset of these resources. Because of the mixed composition of packages consumed by animals, they may often compete for resources which are substitutable, whereas, in contrast, plants may compete for essential resources. Alternatively, if we consider types of foraging, we may expect that motile animals experience consumptive competition whereas plants and sessile animals experience space competition. If such gross generalizations are possible, then details of competitive interactions, and their evolutionary consequences, may be fundamentally different.

1.4 KINDS OF COMPETITION

Resources need not be explicitly considered to recognize different kinds of competition. One could instead consider: (1) the mechanism of the interaction between the competitors; or (2) the kinds of entities interacting. These distinctions could potentially exist whatever the nature of the resources.

1.4.1 Mechanism of the interaction

An alternative approach to classification is to consider mechanisms by which individuals produce the negative effects experienced by other individuals (Miller, 1967). Two basic kinds can be recognized, although these can be further subdivided into as many as six kinds (Schoener, 1983). **Interference competition** occurs when one individual directly affects another. Outright physical attack may occur, or subtler forms of it such as threat behaviour or territoriality. **Exploitation competition** occurs when effects are indirect, and occur solely through reduction of the available pool of resources. The following two examples illustrate these contrasting mechanisms.

Example 1. Exploitation and interference competition in dung beetles

The excrement of large vertebrates provides a rich source of food, exploited by many beetle species. The length of time for which dung remains usable is short, particularly in savannahs where it dries out quickly. Many beetles therefore bury the dung to maintain higher moisture levels. Their egg masses are then deposited on this dung (Crowson, 1981). Such rich patches of resource are rapidly exploited. Bartholomew and Heinrich (1978) describe putting out 1 litre of elephant dung at 10 min after sunset, and collecting 637 beetles attracted to it within the next 30 min. They cite other studies which have found as many as 7000 beetles in one pile of dung. As a consequence, these piles of dung are rapidly depleted, particularly by small beetles which eat it or bury it on site, leaving behind only a thin layer of coarse, inedible fibrous material. Consequently, beetles which require dung for making balls must arrive early, make a ball of dung and roll it away from the site quickly before it is consumed.

Bartholomew and Heinrich (1978) show that success at this exploitation competition can be predicted from body temperature. Beetles with warm bodies have a more rapid rate of dung rolling, which means that warmer beetles are, on average, more successful at exploitation competition. This may be one of the principal advantages of endothermy in these beetles.

The story does not end at exploitation competition. The vast number of individuals harvesting a rich and rapidly vanishing food supply would seem to provide an ideal environment for interference competition as well. Bartholomew and Heinrich note that 'attempted theft of completed dung balls and sustained fighting over partly completed balls are commonplace'. They therefore constructed an arena and explored the interactions between pairs of beetles and artificial dung balls. The winner was the individual which gained access to the dung ball and began rolling it away. They found that the winner was usually the individual with the higher body temperature (Heinrich and Bartholomew, 1979).

These studies not only illustrate the interpretation of exploitation and interference competition, but they suggest that a simple independent variable (body temperature) can be used as a predictor of success.

Example 2.   Interference competition in beetles and flies: The helicopter–gunship strategy

Carcasses also provide a rich supply of food for decomposers. This quality resource is very localized, and is difficult to predict in either time or space. Both flies (Diptera) and beetles (Coleoptera) lay their eggs on such carcasses. Late arrivals are at a severe disadvantage, since not only does the quality of the cadaver decline with time, due to effects of climate and micro-organisms, but it is increasingly likely to be occupied by potential predators and competitors (Crowson, 1981). There are thus many parallels between dung- and carrion-consumers.

In such situations we can postulate that there has been strong selective pressure to locate cadavers early. Evolution of sensory systems and searching ability might be inferred. Invariably, however, these rich resource patches have to be shared. A most interesting example of interference competition occurs under such circumstances.

Carrion beetles (*Necrophorus* spp.) frequently arrive at carcasses already occupied by fly larvae of the genus *Calliphora*. In experimental studies, Springett (1968) showed that when fly larvae are present, the beetles are unable to reproduce successfully on the corpse (Table 1.4). However, under natural

Table 1.4 Interference competition for corpses. The results of experimental cultures using standard corpses (*Apodemus*) inoculated with different combinations of *Calliphora* flies, *Necrophorus* beetles and *Poecilochirus* mites (after Springett, 1968)

Mixture of species on corpse	Number of successful beetle cultures	Number of successful fly cultures
100 fly eggs	–	8
Pair of beetles	8	–
100 fly eggs + pair of beetles	0	8
100 fly eggs + 30 mites	–	0
100 fly eggs + pair of beetles + 30 mites	6	0

conditions these beetles usually carry up to 40 mesostigmatid mites. When the female beetle lands on a carcass, the mites disembark and seek out and kill the fly eggs. The beetles then reproduce successfully. When the larvae pupate, the female abandons the corpse and large numbers of mites depart with her. Other mites leave with the beetle larvae after they pupate. The Coleoptera thus evolved the technique of aerial search-and-destroy tactics millennia before the 'developed' world unleashed it upon peasant villages.

This example illustrates the difficulty in setting limits to the concept of competition. There is a resource in limited supply, and there is good experimental evidence of competitive exclusion in the absence of mites. However, the mechanism of interspecific competition which reverses the species being excluded is the effects of predation by a third party.

1.4.2 Kinds of entities interacting

The most obvious way of classifying competitive interactions is to classify them as those occurring among individuals of the same species (intraspecfic competition) or those occurring among individuals belonging to different species (interspecific competition). This simple classification has dominated studies of competition. Its attractiveness may lie in the species-oriented approach of many ecologists: is competition within the favourite species being studied, or between that species and some other? This is reminiscent of early Western movies, where there good guys in white hats and bad guys in black hats, making the plot relatively easy to follow.

Perhaps a small audience recognizes that characters are not conveniently labelled good and bad, and that characters can fill a variety of roles. The transition from simple dichotomies to multivariate classifications is not an easy one to make, as people well know if they have tried taking parents trained on early Westerns to contemporary films. So it is with classifying competition. Inter- and intraspecific is the classification which has dominated ecology to date, but nature is not a simple cowboy film. Some of the many kinds of competitive interactions we could classify are introduced below. The list is by no means exhaustive.

Intraspecific competition (Fig. 1.4) between pairs of individuals within a species is commonly assessed in plant ecology by sowing seeds at different densities, and recording the performance of individuals (for example, biomass or seed production) at a later date (for example, Harper, 1961, 1977; Watkinson, 1985a; Weiner and Thomas, 1986). It is also measured in replacement series-type experiments (for example, de Wit, 1960; Harper, 1977;

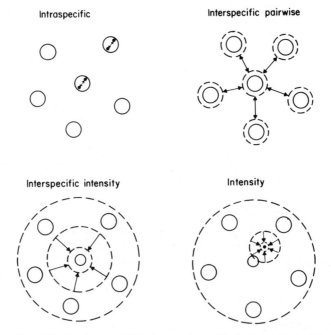

Figure 1.4 Four kinds of competitive interactions. Each circle represents a population; individuals, where necessary, are indicated by solid dots within these circles. The arrows indicate reduction in performance of individuals and populations, resulting from competition. A majority of examples in the literature fall into the top two categories, but in communities populations (left) and individuals (right) can be influenced simultaneously by all neighbours.

Firbank and Watkinson, 1985). In animal ecology such relationships can be assessed by comparing measures of performance and population size collected over many years (Lack, 1966), although if the animals are sufficiently small laboratory experiments (for example, Gause, 1932; Park 1948, 1954; Longstaff, 1976; Gilpin *et al.*, 1986) and field experiments (Connell, 1983; Schoener, 1983) can be used.

Interspecific pairwise competition (Fig. 1.4) is explored when only a single pair of species is examined. It is often compared with a measure of intraspecific competition. This is a popular comparison for both field experiments (Schoener, 1983; Connell, 1983) and laboratory experiments (Gause, 1932; Park, 1948, 1954; Longstaff, 1976; Widden, 1984). A growing number of studies have measured many pairwise interactions simultaneously (Wilbur, 1972; Goldsmith, 1978; Fowler, 1981; Silander and Antonovics, 1982; del Moral, 1983; S. D. Wilson and Keddy, 1986b; Mitchley and Grubb, 1986; Gilpin *et al.*, 1986). One can then ask questions about the relative importance of competition between different populations, or the proportion of possible interactions which are significant.

Competition intensity (Fig. 1.4) refers to the effects of all neighbours upon the performance of a population (left) or individual (right). This can be measured by removing all neighbours and observing the release (if any) of the remaining population or individual relative to unmanipulated control plots (for example, Putwain and Harper, 1970; Fowler, 1981). A possible variant on this is to use transplanted individuals of one, or several species, as a 'bioassay' of the competition intensity in different plots (del Moral, 1983; S. D. Wilson and Keddy, 1981a). Weldon and Slauson (1986) propose that competition intensity can also be measured by comparing the physiological state of organisms in plots with and without neighbours (e.g. Fonteyn and Mahall 1981). Whether to remove intraspecific competition depends upon the question being asked, but the distinction between inter and intraspecific competition may often be necessary. If the growth of a larval anuran is reduced by neighbouring larvae, then its fitness declines with density regardless of whether the neighbours are conspecific or heterospecific (Wilbur, 1972). Similarly, plants will experience nutrient-depletion zones in the presence of neighbours, but the plant may have no way of detecting whether the depletion zone is caused by interspecific or intraspecific neighbours.

Diffuse competition (Fig. 1.5) is closely related to competition intensity. The cumulative effects of neighbours (competition intensity) may range along a continuum of possibilities; on the left, the effects of all neighbouring populations are relatively equal, in which case the competition is said to be diffuse. On the right, one of the neighbouring populations is the primary contributor to competition intensity, and the remaining populations have a minor effect. There is no word currently used to describe this situation in this context; **predominant competition** may, therefore, be a useful term.

Asymmetry and reciprocity (Fig. 1.6). Careless reference to the 'intensity of interspecific competition' between a pair of species implies that each

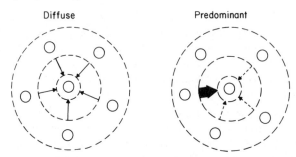

Figure 1.5 Any particular value for the intensity of interspecific competition can be caused by a continuum ranging from diffuse interactions (left) to predominant interactions (right). Experiments are necessary to assess where a particular community falls along this continuum. Intensity and predominance are therefore two independent attributes of communities (symbols as in Fig. 1.4).

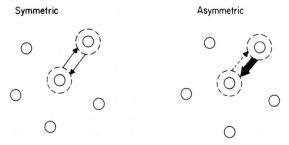

Figure 1.6 Interspecific pairwise interactions fall along a continuum from symmetric (left) to asymmetric (right). On the left each population has an equal effect upon the other, whereas on the right one population is dominant over the other (symbols as in Fig. 1.4). It is therefore inappropriate to write about the 'intensity of competition between two species' without first showing that asymmetric competition is not occurring.

population is having a similar effect upon the other (left), and obscures the possibility that the competitive interactions may be quite unequal (right). These represent two ends of a continuum of possibilities for each possible pairwise interaction in communities. At one end of the continuum are interactions in which the two competing populations are equivalent and are producing equal effects upon each other. At the other end of the continuum one of the populations is so dominant over the other that the effects of the subordinate upon the dominant cannot be detected. The current terminology is confusing. Plant ecologists have used competitive equivalence (left) as their reference point, and therefore use the term 'reciprocal' to describe the former set of conditions (Fowler, 1981; Silander and Antonovics, 1982). In contrast, animal ecologists have used dominance as their reference point (right), and therefore use the term 'asymmetric' to describe the latter set of conditions

(Lawton and Hassel, 1981; Persson, 1985). The terms are therefore both equivalent and opposite. This distinction is illustrated in Fig. 1.6; in the first case, the interaction is symmetric (reciprocal), whereas in the second case the interaction is asymmetric (non-reciprocal).

The above examples do not exhaust the possibilities for recognizing different kinds of competition and competitive interactions. Two others deserve attention and consideration. First, Arthur (1982, 1987) has emphasized competition between different genotypes within populations, and the evolutionary consequences of such interactions. This presents competition within the conceptual framework of evolutionary ecology. Secondly, Buss (1988) has explored competition among different cell lines within individuals, and the implications of this for the evolution of development. Although investigations of intra- and interspecific competition dominate the current literature, future progress may lie along research paths exploring higher levels of organization (e.g. competition intensity gradients) and lower levels of organization (e.g. intra-organismal competition).

1.5 COMPETITIVE DOMINANCE

Competitive dominance is an outcome of interactions where one species suppresses another through exploitation and/or interference competition. It starts with asymmetric (non-reciprocal) competition between individuals, genotypes, or species. The effects of the dominant upon the subordinant are steadily enhanced through two positive-feedback loops (Fig. 1.7). First, there is exploitation competition. The dominant lowers the resource levels for the subordinant, but is simultaneously better able to forage for additional resources itself by reinvesting newly captured resources in further growth. This lowers further the resource levels for the subordinant. Secondly, there is interference competition. The more successful the dominant is at interfering with neighbours, the greater the resources available for supporting further growth of the dominant. This increases further its ability to monopolize resources, increasing both rates of resource acquisition and damage to potential competitors. The relative importance of the two loops is likely to vary from situation to situation, and in some cases effects may be separated into exploitation and interference only with difficulty. However, the end-result is one species suppressing another or excluding it from a given community.

One of the difficulties with discussing dominance arises from the tendency to assume that competitive interactions are symmetric and talk loosely about 'competition' between two species. As soon as there is asymmetric competition, the experience of the dominants and subordinants diverges. It becomes essential to specify whether competition is being viewed from the perspective of the dominant or the subordinant. The analysis of such interactions is clarified by considering that in any competitive interactions there are both effects and responses (Goldberg and Fleetwood, 1987;

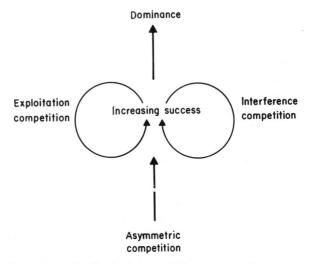

Figure 1.7 The positive-feedback loops which generate dominance. Success at exploitation competition increases the resources available to forage for new resources, and simultaneously reduces the resource supply for neighbours. Increased availability of resources allows some to be channelled to interference competition, damaging neighbours and leaving more resources available for exploitation by the dominant.

Goldberg, 1987). Effects are the negative pressures of each species on the other through exploitation and interference competition; in asymmetric competition the dominant will have most effects. Responses describe the impact which these effects have upon the competitors. The response of a subordinant may be to tolerate the impact of the dominant, in which case it remains present, albeit at a low level. Alternatively, it may escape from the competition by dispersing in space or time to another site (a ruderal or fugitive species). The analysis of asymmetric interactions requires explicit consideration of the effects of the dominant and responses of the subordinant.

These issues are discussed further in Chapter 6, which looks at hierarchically structured communities. It is important here to clarify the distinctions between **competitive dominance** and **dominance**. The word dominant is sometimes used to describe any organism which is abundant in a community. This usage is misleading; abundance need not be the result of competitive dominance. Competitive dominance is abundance achieved as a consequence of exploitation and interference competition for resources – that is, there is an active process of suppressing neighbours (Fig. 1.8, bottom). Grime (1979) describes dominance as a process whereby one species achieves numerical dominance and suppresses others. His use of dominance is not equivalent to the term competitive dominance used here, since Grime includes a second group of effects – a species may become dominant because of inherently better abilities to withstand environmental effects such as fire, infertility or grazing. This

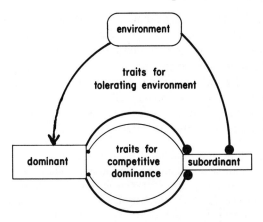

Figure 1.8   The possible interactions between the dominant, the subordinant and the environment. Arrows are positive effects, solid circles are negative effects. Competitive dominance refers solely to the direct links between the dominant and the subordinant (bottom). The environment (top) may independently determine which species dominates a site. In this example the environment is enhancing the effects produced by asymmetric competition, so dominance is only partly attributable to competition.

added group of effects is shown by the upper portion of Fig. 1.8. It seems useful to distinguish between situations where a species is dominant simply because of inherent traits for tolerating the environment and situations where a species is dominant because it has traits for suppressing neighbours. The former type of dominance could occur in the absence of any competition. Only in experiments in which possible dominants are removed and the responses of subordinants observed would it be possible to separate the effects of the two causal agents. In Fig. 1.8 the environmental effects are reinforcing the competitive effects, but it is possible to imagine the opposite situation where the environment weakens the effects of the dominant. In this book competitive dominance is emphasized, but it is important to recognize that both occur in nature and that competitive dominance is a subset of dominance as used by Grime (1979).

Examples of competitive dominance in different plant and animal communities can be found in Chapters 2 and 6. The relationship between dominance and territoriality has been of particular interest to behavioural ecologists. Kaufmann (1983) emphasizes the behavioural basis of dominance in intraspecific competition, noting that 'dominance/subordinance is a relationship between two individuals in which one (the subordinant) defers to the other (the dominant) in contest situations'. It is usually assumed, although rarely shown, that such relationships represent an adaptive compromise for each individual in which benefits and costs of giving in or not giving in are compared. Dominance provides priority of access to resources, and is therefore a behavioural aspect of asymmetric interference competition. E. O. Wilson

(1975, 1978) has reminded humans that ecological and evolutionary principles apply to humans as well as any other organism. Dominance has been explored in human pairwise interactions (for example, Maccoby and Jacklin, 1974). The mechanisms of dominance are often easiest to observe in human interactions. Some of the most dramatic examples of dominance are found where human societies compete for access to global resources. These include socio-economic positive-feedback loops and military interference competition. Examples could come from China in Tibet or France in Algeria, but Table 1.5 illustrates

Table 1.5 Interference competition for global resources, illustrated by dominance and asymmetric competition among humans. This example illustrates some measures of dominance (top) and mechanisms of interference competition (bottom) from one of many possible global conflicts. The mechanisms of interference competition are easily observed in human interactions, and the asymmetry of this interaction is obvious upon inspection (data from Matthews and Morrow, 1985; Melrose, 1985; Oxfam Canada; school category includes adult education centres)

	USA		*Nicaragua*
Index of dominance			
Gross national product			
(US $ *per capita*)	14 080		884
Area (km²)	9 363 000		148 000
Population (millions)	241		3
Interference competition			
American civilians		Nicaraguan civilians	
killed by Nicaragua	0	killed by USA	> 7500
American teachers		Nicaraguan teachers	
killed or kidnapped		killed or kidnapped	
by Nicaragua	0	by USA	> 300
American schools		Nicaraguan schools	
closed by		closed by	
Nicaragua	0	USA	> 1000
American health		Nicaraguan health	
centres eliminated		centres eliminated	
by Nicaragua	0	by USA	> 40

the USA in Latin America. The USA has a dominant economy dependent upon the acquisition of resources from Third World countries (Lappe and Collins, 1982; Myers, 1985). As success at acquiring resources increases, large corporations have the wealth to buy more land and build more mines in developing countries, further increasing rates of resource flow into the dominant economy. This further reduces resource supplies to the local economies (Lappe and Collins, 1982), increasing economic dominance and inequities further. Resource depletion can produce hunger in countries that have more than sufficient agricultural land to feed their own populations. This is straightforward exploitation competition. The resource surplus of the dominant can then be used for interference competition, where military dictatorships that are favourable to resource extraction are installed and maintained by military and economic aid (Chomsky and Herman, 1979; Klare and Arnson, 1981). Assassinations by death squads and the destruction of hospitals, co-operative farms and day-care centres are part of the interference feedback loop.

1.6 BACK TO BASICS

The title of this section is not an appeal for more of the same, but rather a suggestion that fundamental progress is likely to come from re-evaluating fundamental assumptions. The overview in this chapter illustrated the many kinds of competition that can exist in nature. Important generalizations about the nature of competition may emerge when we find a classification of resources or classification of competitive interactions which has high explanatory and predictive power. To succeed, there will have to be objective (measurable) criteria for assigning situations to the different categories (or sections of a continuum). This is particularly true when it comes to measuring resource characteristics.

Important progress is possible here if innovative approaches are tried more often. That pairwise interspecific and intraspecific competition has received such attention is remarkable, given the range of possibilities which can be postulated to exist.

The next three chapters (Chapters 2–4) explore how the conceptual foundation in Chapter 1 has been used to explore nature. Chapters 5–7 present some avenues where rapid progress appears possible. Chapter 8 reconnects to Chapter 1 by going back to basics and asking why ecologists do science in the way in which they do, and what this has to do with progress in the study of competition.

1.7 QUESTIONS FOR DISCUSSION

1. What are the strengths and weaknesses of the classifications of resource types presented? Can you devise an alternative classification of resource types?

2. How do we make such classifications operational? That is, how would we actually measure them in order to make ecological predictions?

3. How would you design an experiment to measure each of the kinds of competition described?

4. Are there other kinds of competition you believe should be recognized? Why? How would you measure them?

5. Are there any reasons for expecting the biosphere to be structured primarily by a subset of the above possibilities?

6. Why are studies of inter- an intraspecific competition so abundant in the ecological literature?

7. What would be the ecological and evolutionary consequences of asymmetric competitive interactions between pairs of populations?

8. Can you find examples of each kind of competition in the current literature?

9. Can you argue from first principles which kinds of competition should predominate in each of the five kingdoms of living organisms?

10. Weldon and Slauson (1986) have defined competition as 'the induction of strain in one organism as a result of the use, defense, or sequestering of resource items by another organism'. This follows Levitt (1980) who uses the term strain to describe an organism's response to environmental stress. Compare and contrast their definition with that on page 2.

2 Competition in action

Ought we, for instance, to begin by discussing each
separate species – man, lion, ox, and the like – taking
each kind in hand independently of the rest, or ought
we rather to deal first with the attributes which they
have in common in virtue of some common element of
their nature, and proceed from this as a basis for the
consideration of them separately?

Aristotle, De Partibus Animalium

... all the laws of nature and all the operations of
bodies without exception are known only by
experience....

D. Hume, Enquiry Concerning Human Understanding

And though I fret and worry till I'm weary,
When? How? and *Where?* remains the fatal query;

Mephistopheles, in J. W. Goethe, Faust: Part 2

It is one thing to define competition and create a vocabulary, but quite another
to test whether these terms are in fact useful. Is there evidence for the existence
of competition in real communities of organisms? Is it of broad general
occurrence, or only present in isolated and special situations? Do existing
examples fall usefully into categories such as interference or exploitation
competition? Do different groups of organisms differ in the degree to which
competition controls population sizes?

There are two ways in which we might go about addressing such questions.
First, we might begin by asking what all living systems have in common. This
naturally leads to a consideration of the thermodynamic properties of living
organisms. Beginning from such generality, we could then look for differences
among kinds of situations (habitats, resources and organisms and approach the
more complex real world. Alternatively, we might begin with details by going
out to real ecosystems and measuring some of the properties described in
Chapter 1. Having done this for enough examples of different systems, we might
then look for general principles. In either case, the objective is to determine

which parts of the conceptual foundation in Chapter 1 are appropriate for building models and general theories about the role of competition in nature.

The latter approach of collecting special cases is more common. This chapter tries both approaches. It first reviews thermodynamic models which suggest that competition is likely to be a force that is present in any large assemblage of molecules through which energy is flowing. It then explores selected case studies to illustrate some of the species and habitats which have been used to study competition. Finally, it addresses the problems which arise in trying to draw inferences about general principles from collections of special cases. These examples provide a context for chapter 3, 'Modelling Competition', by supplying concrete examples of competition in action in real ecosystems. Chapter 4 then addresses the broader scientific issues which arise when planning research strategies to probe nature.

2.1 COMPETITION AND THERMODYNAMICS: BASIC PRINCIPLES

Morowitz (1968) explores biological and ecological processes assuming that 'biology is a manifestation of the laws of physics and chemistry operating in the appropriate system under the appropriate constraints'. This section explores only a small fraction of his ideas, primarily those providing a context within which we can view competition. Morowitz notes that living systems are at a high potential energy level; i.e. the living state has a very unlikely distribution of covalent bonds compared with the equivalent equilibrium state at either the same total energy or the same temperature. Living systems, he shows, are not at thermodynamic equilibrium, nor could they have spontaneously originated from a chance reaction near thermodynamic equilibrium. The Earth, however, is not an equilibrium system, but a steady-state system with a steady flow of energy as sunlight flows from a source (the Sun) to a sink (space). Morowitz shows that this energy flow organizes matter and produces systems with high potential energy.

The origin of life is therefore an inevitable consequence of physical laws, and not a chance event. He then explores 13 biological generalizations based on equilibrium organic chemistry. The general theme of these is that

> the tendency to organize is a very general property of a certain class of physical systems and is not specifically dependent on living processes. Molecular organization and material cycles need not be viewed as uniquely biological characteristics; they are general features of all energy flow systems. Rather than being properties of biological systems, they are properties of the environmental matrix in which biological systems live and flourish.

The first few of his 13 generalizations provide the foundation for a conceptual model relevant to the origin of competition. Living systems on Earth are primarily water; water is an important metabolite as well as a solvent. Within this aqueous system, the major atomic components are carbon, hydrogen, nitrogen, oxygen, phosphorus and sulphur, with a variety of minor and microconstituents (Table 1.1). Imagine a simple mixture of CHNOPS molecules (e.g. H_2O, CO_2, N_2, NH_3 and CH_4). If energy flows through this system so as to raise the average potential energy, what will be the distribution of chemical species? Compounds in higher energy states will increase at the expense of the abundant low-energy compounds. There is no alternative; if the energy is supplied in a form such that it goes into chemical bond energy, then rearrangements must occur, leading to different bonds and different molecules. This process is observed in experiments which explore the kinds of biological molecules produced in environments simulating the prebiotic period on Earth (Orgel, 1973).

Once a pool of molecules of slightly higher potential energy is created by energy flow, interactions occur within this pool, leading to the creation of slightly larger molecules with still-higher potential energy. As long as energy flow is maintained, one can envisage a pyramid, with pools at different potential energy levels, each pool serving as raw material for the pool above it, and each similarly using the pool below it as a source of raw material (Fig. 2.1). If the energy flow is turned off, the system naturally gradually collapses to states with lower potential energy, but this possibility need not be pursued here.

Using this foundation, consider one potential energy level in Fig. 2.1. At this level are molecules with similar potential energy, constantly formed from a pool of lower-energy molecules (a resource pool) and occasionally converted into molecules of higher potential energy.

Within such potential energy pools a primitive form of natural selection is taking place. Those molecular forms which are 'unstable' by definition break apart into lower-energy molecules which then are returned to the resource pool. Thus, certain kinds of molecules proliferate at the expense of others. Their abundances are determined by three factors: (1) the rate at which molecules from the lower resource pool are converted to higher-energy molecules; (2) the rate of decay of the higher-energy molecules to the lower-energy ones (their 'stability'); and (3) the rate at which these higher-energy molecules react to produce ones of even higher potential energy.

It is clear that those molecular forms which proliferate will have three properties. First, they will be rapidly formed from lower-energy molecules; if they catalyse such interactions, this process will be enhanced. Thus, one can draw the analogy of consumer molecules, each dependent upon the conversion of molecules from low to high potential energy. Secondly, the proliferating molecules will have traits which increase persistence through time. Molecular stability is an initial prerequisite, but cell walls can be seen as one relatively simple method of further enhancing stability. Thirdly, those molecules which resist conversion to higher-energy molecules will also accumulate. Cell walls could also accomplish this.

At this molecular level it is already possible to discern the essential processes

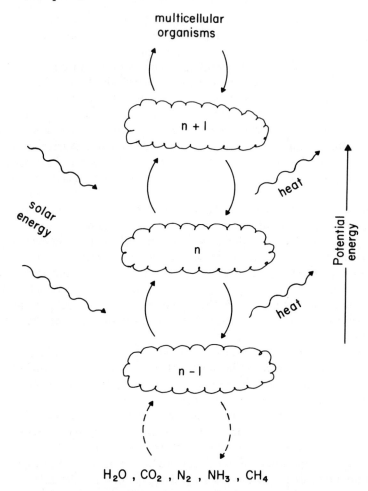

Figure 2.1 Solar energy creates high-energy molecules out of simpler low-energy molecules. Complex molecules and multicellular organisms are inevitable thermodynamic consequences of energy flow in the biosphere (Morowitz, 1968). For any arbitrary level of potential energy there is a restricted pool of substrate molecules at the next-lower level, so that even in a simple molecular system a form of resource competition can be observed.

with which ecologists are concerned. Resource pools are consumed and converted into more-organized (higher potential energy) systems. The abundance of the higher potential energy systems like protein is limited by the abundance of the resources like NO_3, and the rate at which they can be 'harvested'. There is, therefore, a form of intermolecular competition for the resource pool. With sufficient imagination, predator–prey interactions can also be seen, with the higher-energy molecules preying upon the lower-energy ones.

There are therefore good thermodynamic reasons for expecting competition in nature. They are based on the observation that the accumulation of more-complex molecules is limited by, among other factors, the pool of resources. The variety of life-forms is staggering, and it is so easy to get caught up on fascinating details of form, function and natural history. However, if the systems were stripped of all detail, they would look very much like the thermodynamic model that Morowitz presented for the behaviour of simple chemical systems.

Of course, even if this is true, it does not prove that competition is universal. In fact, the seeds of two major counter-arrangements are hidden within that same molecular model. If the energy flow fluctuates, then the amount of a particular compound may be less dependent upon the resource pool than upon the time since the last perturbation. Thus, we have the argument that competition is not important if systems are rarely near equilibrium (that is, in Morowitz's terms, if there is variation in the energy flow producing the steady state). If molecules are continually converted to other forms by enzymes ('preyed upon'), then their abundance may be set as much by the rate of removal from the pool than by their rate of production. Thus, we have the argument that predation reduces competition.

2.2 CASE STUDIES OF INTRASPECIFIC COMPETITION

Instead of arguing from first principles, we could try to apply the definitions in Chapter 1 and measure different aspects of competition in natural systems. In Chapter 4 we explore the different conceptual approaches to studying competition in nature. Here we simply examine some specific examples where experiments have been used to study aspects of competition. Case studies such as these provide useful reference points for the next two chapters, as well as illustrating the use of definitions presented in Chapter 1.

One of the simplest methods for detecting intraspecific competition is to test for a negative relationship between performance and population density. Performance (or fitness) can be measured in many ways, depending upon the organism and the circumstances – growth rates, survival rates and reproductive output are common examples. If the correlation is negative (higher density leading to lower performance) there is evidence for intraspecific competition. Figure 2.2 illustrates possible relationships between performance and density, emphasizing that for any species there may be a different performance–density relationship for each habitat. Also, only a portion of the line may be found in nature; experimental manipulation may be necessary to observe the hatched sections. Figure 2.3 shows, for a species of bird (Great tits), how two different measures of performance decline with density. The top illustrates the evidence for intraspecific competition of a very specific kind, intersibling competition among young of the same parents, by showing that the greater the number of young in a brood is, the smaller the mean weight of nestlings. The bottom

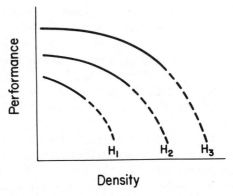

Density

Figure 2.2 Hypothetical relationships between performance and density, which are consistent with intraspecific competition. Measures of performance may include survival, fecundity or size. Each line represents a different habitat, emphasizing that performance–density relationships may differ with environmental conditions. In this case both the intercepts and shapes of the curves vary among habitats (see Keddy, 1981, for other possible situations). The hatched sections of the lines represent situations not normally found in nature, which can only be observed by experimentally increasing population density.

shows that, for the population as a whole, the mean clutch size (number of eggs in a nest) declines with increasing population density.

The relationship between the size of individuals and density has been extensively studied in plant populations, partly because of the obvious agricultural implications. There is a well-established relationship between the mean mass of individuals and the density at which they are grown (Yoda *et al.*, 1963; Harper, 1977; Westoby, 1984), with a slope of $-3/2$. Much of the data for this relationship comes from the experimental study of plant monocultures, but Gorham (1979) has shown that there is a strong relationship between shoot weight and shoot density across 29 different plant species (Fig. 2.4) ranging in size from trees (upper left) to a moss (lower right).

Such correlational approaches could be applied to most kinds of organisms, although, as Fig. 2.3 suggests, many years of hard work may be necessary to accumulate the data. At the same time this approach has three weaknesses which, depending upon the particular system, may be fatal flaws. The first, and most obvious, is that correlation does not demonstrate cause and effect. Since different habitats, years, nests or quadrats provide each datum point, it is possible that the correlation is spurious. If, for example, habitats providing the most food were also most exposed to predation, high individual performance and low density could be found correlated in the complete absence of interspecific competition. Similarly, in plant communities plant performance will generally by positively correlated with the available soil resources (Harper, 1977; Chapin, 1980; Tilman, 1982); however, if soil pathogen activity

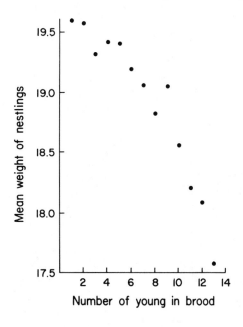

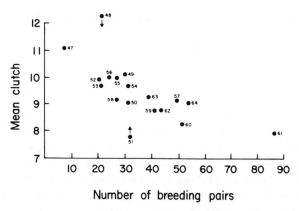

Figure 2.3   Two relationships suggesting intraspecific competition in Great tits. The first shows that nestlings weigh less as the brood size increases. The second shows that the mean size of clutches (eggs per nest) declines with increasing population density (each point represents a year from 1947 to 1964). (After Lack, 1966.)

(Burdon, 1982) was also positively correlated with soil fertility, then the combination of increased mortality and higher resource levels in the more fertile habitats could produce a negative correlation between performance of survivors and population density. The correlation might be interpreted as intraspecific competition, but would really be the result of the positive correlation between soil fertility and pathogens.

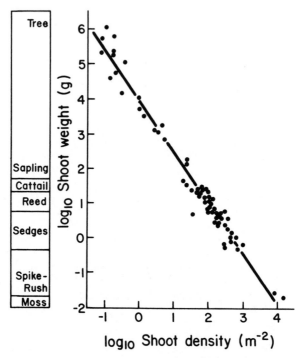

Figure 2.4 The relationship between performance (shoot weight) and density for 65 stands of plants representing 29 species from trees (upper left) to a moss (lower right). Intraspecific competition appears to set an upper limit to the number of shoots of a specific size which can coexist. (After Gorham, 1979.)

A second, related, weakness is that such studies assume that only population density measures are needed to describe a particular habitat. Given the range of habitats which most species occupy, the performance–density relationship is likely to differ among them. Species may have habitats where populations have density-dependent relationships, and others where these are absent. Most species will therefore be represented by a family of performance–density response curves (Keddy, 1981).

A third potential problem is that a species may have the same density in two habitats, but in one there may be a positive population growth rate and in the other a negative growth rate. The equivalence of density may therefore be transitory and misleading.

These criticisms can be overcome only by changes in approach to such studies. The first is that densities need to be experimentally manipulated so that the observer knows that the data points differ in population density with other environmental effects randomly distributed. Secondly, different habitats or populations need to be compared to determine whether the intensity of intraspecific competition varies among habitats or populations. Thirdly, population growth rates need to be measured. This takes considerably more work, so there are few examples in the literature.

2.2.1 Density-dependence in annual plants

The first two of the above points were addressed in a field study of an annual plant, *Cakile edentula*, which grows on sand dunes along the coast (Keddy, 1981, 1982) and the third in a re-analysis of these data (Watkinson, 1985b). *Cakile edentula* plants can be found in a wide range of habitats and show corresponding changes in plant size, reproductive output, survival and population density (Keddy, 1981, 1982). These habitats are arranged along a gradient. At one end one can find large plants with thousands of seeds growing amidst decaying seaweeds on open sand beaches. At the other end tiny plants with but one or two seeds can be found beneath a canopy of dune grasses.

Density dependence was tested for by sowing a range of seed densities, allowing germination and growth to occur for one summer, and then testing whether performance was negatively correlated with sowing density. Two measures of performance were used: percentage of seeds sown which produced reproductive plants, and mean number of fruits produced per plant.

The principal results are shown in Fig. 2.5. Density-dependence clearly varied among the three habitats. In the middle of the gradient there was no evidence for it. At the seaward extreme crowding significantly reduced only reproductive output, whereas at the landward end both reproductive output and survival declined with density. Also, in the landward habitat, the predominant effects of crowding were the reduction of survival, whereas at the seaward end it was reduction in reproduction.

There are two general conclusions for studies of intraspecific competition. First, the relationship between performance and density is not a trait of a species, or a population alone, but is strongly dependent upon the environment itself. Second, the dependent variable selected is extremely important: if only survival or only reproductive output had been used, then entirely different conclusions would have been reached about which habitat produced the most intense intraspecific competition.

The problem with this experimental approach is that it does not yield unequivocal statements about the actual intensity of interspecific competition in real populations unless the usual range of population density in the habitat is known. If, for example, seaward populations usually occurred at low densities (and they do), then the intensity of interspecific competition shown in the seaward site is a potential which is rarely realized. Thus, such experimental data need to be combined with measures of population density.

One problem of interpretation still has not been removed: how are we to know that the density dependence is attributable to intraspecific competition? Many potentially confounding effects of temporal or spatial variation have been eliminated by using experimentally produced densities, so a large number

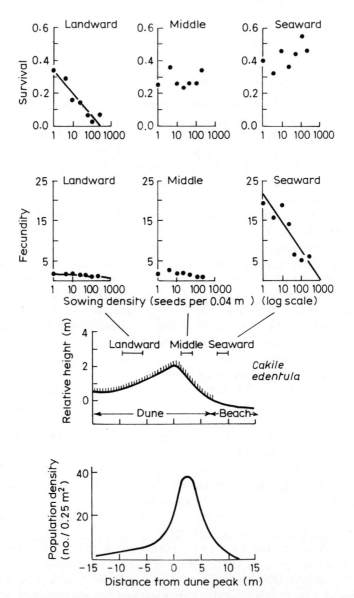

Figure 2.5 A study of an annual plant growing on sand dunes, which shows that the effects of density upon survival and fecundity vary among habitats. The two measures of performance – survival and reproductive output – had very different responses to habitat and density. Thus, the habitat and the performance measure selected may determine whether a researcher detects competition in the field. For example, if the study had been conducted measuring only reproductive output in the middle section of the dune, which is where most of the plants are found, then no evidence for competition would have been detected. (After Keddy, 1981, 1982; Silvertown 1987.)

of potential alternative hypotheses have been eliminated. However, others remain. If predation were density dependent, then the density dependence of survival could be attributed to predation, not competition. In fact, epidemics of damping-off disease do occur in *C. edentula* populations, although there was no evidence for it in the year in which this study was conducted. Mixed strategies can even be imagined where intraspecific competition weakens individuals which then fall prey to pathogens, or are buried by drifting sand because of their small stature. One source of evidence for competition would be to augment the supply of a resource that is postulated to be limiting. By running a series of treatments with nitrogen fertilization, Keddy (1981) showed that reproductive output of low-density landward plants (but not their survival) increased when this potential resource was supplied. This is additional evidence that intraspecific competition for nitrogen limited plant size. Since there were no effects on survival, we may postulate that an independent factor (perhaps competition for another resource, or predation) controlled survival.

The greatest weakness in this sort of experimental study is the absence of information on year-to-year variation. If storms destroy seaward populations in most years, then intraspecific competition may be much less important than the data in Fig. 2.5 suggest. This criticism could only be answered by repeating the study in several years with very different weather.

2.2.2 Sperm competition: a driving force of intraspecific competition

The definition of competition included the term 'resources'. In intraspecific competition resources are normally considered to be raw materials necessary for the growth and reproduction of populations. However, the acquisition of resources is only one step towards eventual reproduction; mates can be viewed as the ultimate limiting resource for intraspecific competition. Competition among sperm (and among the individuals producing them) for access to eggs may therefore be one of the most intense and widespread intraspecific competitive interactions in the biosphere.

Competition for mates has a basic asymmetry which is a direct result of anisogamy (difference in sizes of gametes) – there are vast numbers of sperm produced by males relative to the limited number of eggs produced by females. Among anisogamous taxa we can recognize a further subdivision, into groups with and without internal fertilization. The evolutionary strategies which maximize reproduction for particular males or females will clearly differ between these two cases. The majority of studies of sperm competition have been carried out among taxa with internal fertilization, so these are emphasized here.

Both exploitation and interference competition can be recognized. Exploitation competition can be invoked as the selective force acting upon sperm anatomy and morpholgy to maximize success in the race up reproductive tracts to locate and fuse with eggs. Penis morpholgy and size can be seen as the result of selection to deposit sperm as close as possible to eggs (Smith, 1984). The chemical composition of seminal fluid may provide sperm with nutrition and induce uterine contractions in females to assist sperm movement (Smith, 1984). Mechanisms of interference competition are the most exquisitely developed and documented. Consider the following possibilities. Males may secrete plugs to block the female tract and prevent later successful matings; this is found in acanthocephalan worms, insects, spiders, mammals and snakes (G. A. Parker, 1984). The penis may play a dual function; in addition to introducing sperm it can be used as a scoop to remove sperm deposited during earlier matings (Waage, 1979, 1984). The male may prevent other males from gaining access to the female tract while his sperm swim towards the eggs; prolonged copulation and post-copulatory guarding are known from many animal groups. Lastly, hiding during copulation ('take-over avoidance') can prevent other males from finding the pair and interrupting mating or later adding their sperm to the female tract (G. A. Parker, 1984).

Like much of evolutionary ecology, such work can be criticized as *post hoc* explanations rather than critical tests of hypotheses. However, sperm competition can be studied experimentally. Genetic markers can be used to study the success at fertilization of sperm from different males in sequential matings, and to study specific questions asked about advantages accruing to the last male to mate with a female. Similarly, mutant males lacking certain behaviours can be compared with normal males to measure the costs and benefits of different male reproductive strategies.

Mates and eggs can be viewed as ultimate limiting resources, and the intensity of intraspecific competition resulting is illustrated by the large number of examples of interference competition for mates.

2.2.3 Interference competition: nest-destruction by wrens

One of the advantages of studying conspicuous organisms with easily visible behaviour is that, unlike with plants or sperm cells, it may be easier to see the mechanisms of competition. A conspicuous example of this is egg- and nest-destroying behaviour seen in some passerine birds. Many species of wrens, for example, will peck eggs as well as destroy the nestlings of neighbouring individuals. It seems unreasonable to classify this behaviour as predation, since the egg contents are often not eaten, and marsh wrens are not equipped to eat the flesh of the nestlings they kill. The alternative hypothesis is that nest-destroying behaviour has evolved as a mechanism of interference competition.

The interactions between wrens and blackbirds have been explored by Picman (1980, 1984). If nest-destroying behaviour is a mechanism of interference competition, we would predict it should occur with higher frequency at higher population density. In the case of wrens nesting with blackbirds, however, the problem is more complex because there is the possibility of nest destruction occurring for both inter- and intraspecific competition, depending upon the relative abundance of wrens and blackbirds in a marsh.

Picman (1984) studied marsh wrens and red-winged blackbirds in deltaic marshes along the coast of British Columbia, Canada. For 5 years he monitored the densities of nesting males of each species. Each year the nest-destroying behaviour of wrens was assessed as being more strongly inter- or intraspecific. This was done by collecting nests of both species, attaching them to a stake, and placing pairs of nests within the territories of male wrens when they were absent. When the male wren returned, observers recorded whether it explored the experimental wren or blackbird nest first. Since the two types of nests are very different in size and shape, this test was designed to measure whether male wrens were more disturbed by nests of conspecifics or by those of blackbirds. Figure 2.6 shows the relationship between the type of nest first visited and the population density of wrens.

At low wren densities, wrens preferentially visit red-winged blackbird nests. Since it is presumed that in real nests this would be followed by attacks on eggs or nestlings, it appears that this behaviour is directed at other species when wren densities are low. As wren population densities increase, however, greater emphasis is apparently placed on visiting (and presumably attacking) the nests of conspecifics. This pattern is primarily due to changes in wren density over

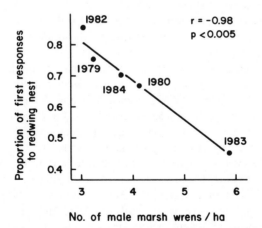

Figure 2.6   Nest-destroying behaviour of wrens may function as a means of both intra- and interspecific interference competition. The relative importance of intra- and interspecific competition was measured by allowing territorial wrens the opportunity to explore both a redwinged blackbird nest and a conspecific marsh wren nest. Wrens increasingly explore the nests of conspecific males as wren density increases. (After Picman, 1984).

the years: over the duration of this study, male red-winged blackbird densities stayed nearly constant at 1.5 males ha^{-1}, whereas population densities of wrens fluctuated between 3 and 6 males ha^{-1}.

This study this illustrates an obvious form of interference competition, and strongly suggests that the relative importance of intra- and interspecific competition varies between years. As with any comparative study, we have to assume that variation in the population densities of wrens were not confounded by changes in other factors which could influence behaviour. The alternative approach – experimentally manipulating the densities of both species – is not really feasible in this system.

2.3   CASE STUDIES OF INTERSPECIFIC COMPETITION

The dung beetle and carrion beetle examples in Chapter 1 described two cases of interspecific competition among insects for high-quality resource patches. The following examples present four different and complementary field studies on interspecific competition.

2.3.1   Competition among grassland plants

Fowler (1981) studied competition in vegetation on a section of the campus of Duke University, North Carolina. Although the field contained more than 50 species, she concentrated on 12 of the more common ones. Three basic questions were asked. Is there evidence for competition? What proportion of the possible pairwise interactions are significant? Is this competition diffuse, or do only a few of the pairwise interactions predominate?

All species manipulated were perennial herbaceous species; they include grasses with varying capacity for vegetative spread, ranging from none (e.g. *Anthoxanthum odoratum*) to stolons (*Cynodon dactylon*) and rhizomes (*Poa pratensis*), and two forbs (*Plantago lanceolata* and *Rumex acetosella*). The design was based on removing selected populations, and testing whether other populations responded to these removals. The removals were carried out in 1975 and the effects measured 2 years later (Table 2.1).

In general, the effects of pairwise removals were relatively small, and only 9 of the 50 pairwise effects tested in Table 2.1 were significant. These significant increases were all of similar magnitude. Fowler concludes 'this is a community characterized by relatively weak and approximately equal competitive relationships among all its competent species'.

The removal of groups of species produced somewhat larger effects, as shown by the columns where 'all grasses' or "all dictotyledons' were removed. Where all plants were removed, only two species, *Plantago lanceolata* and *Rumex acetosella*, responded significantly.

Table 2.1 also permits assessment of competitive asymmetry. Fowler found

Table 2.1 The effects of pairwise removals of plants from abandoned land in North Carolina (*$P < 0.55$; **$P < 0.01$; after Fowler, 1981)

Species responding	Removals							
	Pl	Pd	Pl	Cd	Cc	g[a]	d[b]	all[c]
Plantago lanceolata		–	–	–	–	*	**	**
Paspalum dilatatum	*		–	*	–	–	–	–
P. laeve	–	–		–	–	*	–	–
Cynodon dactylon	*	–	–		–	*	**	–
Carex cephalophora	–	–	*	–		–	*	–
P. ciliatifolium	–	–	–	–	–	–	–	–
Setaria lutescens	–	**	–	–	–	*	–	–
Poa pratensis	–	–	–	*	**	–	–	–
Rumex acetosella	–	*	–	–	–	*	–	**
Salvia lyrata	–	–	**	–	–	–	–	–

[a] All grasses removed; [b] all dicotyledons removed; [c] all plants removed.

no significant correlation between reciprocal pairs of removal responses, concluding that competition was asymmetric.

It is not clear how far we can generalize from this study. For one thing, mowing to a height of 5 cm was continued throughout the experiment, and this would almost certainly have the effect of reducing the magnitude of competitive interactions. Also, considering the range of morphological variation in many plant communities, this one was made up of relatively similar species. Since the vegetation consisted of many alien species including some from Europe (e.g. *Anthoxanthum odoratum*) or South America (*Paspalum dilatatum*), it is not clear whether native plant communities would have similar attributes. In spite of such limitations, this study is an excellent example of the sort of knowledge which can be gained only by systematically removing populations from communities.

2.3.2 Competition among salamanders in ponds

Wilbur (1972) explored competition in a community of North American amphibians which breed in temporary ponds in the spring. This community consisted of three species of mole salamanders (*Ambystoma* spp.). These species return to water to breed, but normally lead terrestrial lives as adults. They were part of a larger amphibian community including Tiger Salamanders, American Toads, Grey Tree Frogs and Wood Frogs. Predators included other amphibians, insects, leeches and birds.

Wilbur placed cages in the shallow water, and inoculated each cage with different numbers and kinds of amphibian eggs. The experimental design included single-species, two-species and three-species systems. At the end of the summer, Wilbur measured three dependent variables: survivorship, body weight and length of time of the larval period for all survivors. The latter dependent variable would be particularly important in years when ponds dry out entirely.

The experiment showed that there was intense interspecific competition, although the details varied with the particular species and the dependent variable. Figure 2.7 compares body weight at metamorphosis for *Ambystoma laterale* and *A. tremblayi*. The addition of 32 individuals to the 32 in the single-species pens reduced body weight in *A. laterale*, but had only a negligible effect on *A. tremblayi*; thus *A. laterale* is apparently much more sensitive to increased density. For both species, adding conspecific individuals had similar effects to adding individuals of other species; that is, intraspecific competition was equal to (or possibly slightly less than) interspecific competition. Given the extreme similarity of these species, this is perhaps not surprising. Note that further increasing density by adding 32 of each of the two other species led to increased body size in *A. laterale* but decreased body size in *A. tremblayi*. Perhaps one might speculate that *A. laterale* larvae were killed by the presence of competitors, thereby allowing the body size of the survivors to increase. However, survival of *A. laterale* in single-species cages was 33%, and in the

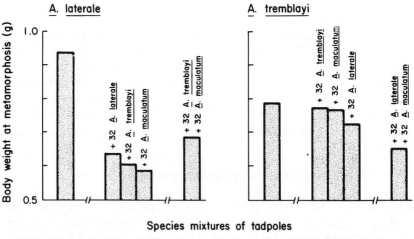

Figure 2.7 Body weight at metamorphosis in the salamanders *Ambystoma laterale* and *A. tremblayi* when grown in pens containing 32 individuals (control), and with experimental additions of either more conspecifics or two other salamander species. There is no evidence that intraspecific competition is greater than interspecific competition. (Data from Wilbur, 1972.)

three species cages was 34%, so differential survivorship is not the explanation for the apparent positive response to increased density.

Wilbur also asked whether the three-species system had emergent properties which could not be predicted from the separate analyses of subsystems. Two results only are discussed here. The factorial design which Wilbur used allows the construction of trend surfaces for the responses of each species to changes in population density. Figure 2.8(a) shows the trend surface for *A. laterale*. Adding 32 of either conspecific produced a sharp drop in body weight, but adding 64 of either made little extra difference. Thus, performance was not simply a linear function of density. When the 64 individuals added consisted of 32 of each species, a further drop occurred. A mixture of two species had more effect than the same number of either one competitor or the other. The performance of *A. laterale* could not be predicted by simply adding effects of densities or species; in contrast; the trend surface for *A. tremblayi* is quite smooth, indicating simple additive effects (Fig. 2.8(b)).

2.3.3   Competition among desert granivores

Many ecologists look for pairs of similar species to conduct removal experiments and test for the effects of competition between them. The following study illustrates a different approach – the removal of entire groups of species from a community built around one resource: seeds.

In deserts seeds constitute a major resource for a taxonomically diverse array of consumers including rodents, ants and birds. In a series of studies, Brown *et al.* (1979, 1986) have conducted removal experiments from a study area in the Sonoran Desert.

This community has three groups of consumers which feed on seeds; a majority of the seeds are produced by annual plants. The first class of consumers is nocturnal rodents which are resident for the entire year. Most

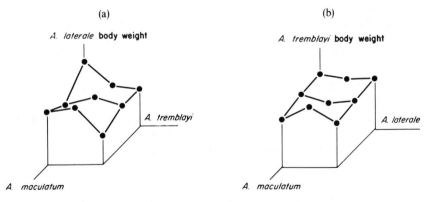

(a)   (b)

Figure 2.8   Body weight at metamorphosis for two salamander species plotted against densities of neighbours in experimental salamander communities. (After Wilbur, 1972.)

collect seeds and store them underground for use in times of food shortage. These include kangaroo rats (*Dipodomys* spp.), pocket mice (*Perognathus* spp.) as well as the more omnivorous deer mice (*Peromyscus* spp.) and harvest mice (*Reithrodontomys* spp.). The second group of consumers is ants. Some are specialized seed eaters (e.g. *Pogonomyrmex* spp. and *Veromessor* spp.) and some are more omnivorous (e.g. *Novomessor* spp. and *Solenopsis* spp.). They live in colonies numbering to the tens of thousands. Workers collect seeds and store them in underground galleries for use in periods of food shortage. The last major group of consumers is birds, particularly sparrows, which can travel over large distances to exploit seeds in areas of temporary abundance. By excluding one or more of these groups from test plots and testing whether the others respond, it is possible to measure the competitive interactions within and among these groups.

One set of experiments was designed to investigate the interactions between ants and rodents (Brown and Davidson, 1977). Eight circular arenas, each with an area of 0.1 ha, were fenced off from the surrounding desert. There were two replicates of four treatments: ants excluded, rodents excluded, both rodents and ants excluded, and neither rodents nor ants excluded. The rodents were excluded by wire-mesh fences and any inside were removed by trapping. Ant colonies were eliminated by localized treatments with insecticide. Compared with control plots, the ants increased where the rodents were removed, and the rodents increased where the ants were removed (Table 2.2). Two years later the resource supply was measured, and showed that seed density and plant density

Table 2.2 Effects of experimental removals of seed-eating ants and rodents from fenced-in plots in the Sonoran Desert (after Brown and Davidson, 1977)

	Controls	Rodents removed	Ants removed	Percentage increase
Ant colonies	318	543	–	71
Number of rodents	126	–	151	20
Biomass of rodents (kg)	4.2	–	5.4	29

increased in the removal plots (Fig. 2.9). The table also shows that the competition between ants and rodents was very asymmetric: the removal of rodents led to a 71% increase in the number of ant colonies, whereas the removal of the ants led to only a 20% increase in the number of rodents.

This study illustrates that although ants and rodents are very different kinds of organisms, they are apparently competing for a common resource. Considering only interactions between species of rodents or between species of ants would have provided an inaccurate picture of the degree to which competition for seeds controls the abundance of these species.

Because the study was conducted in a single area of the Sonoran Desert, we

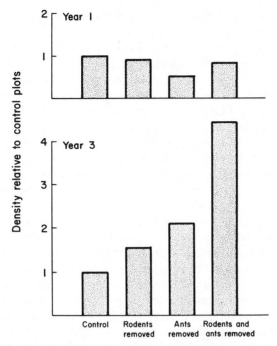

Figure 2.9   Effects of excluding rodents and ants on the density of seeds in the soil. This illustrates changes in resource levels after the removal of possible competitors. (After Brown *et al.*, 1986, Fig. 3.3).

cannot know how general this phenomenon is. A similar set of experiments was later conducted in the Chihauhuan Desert, and the results are far less dramatic. Exclusion of ants appeared to have no effects on rodents, and the effects of rodents on ants are also difficult to detect. Brown *et al.* (1986) suggest that this can be attributed to differences between the sites, particularly in climate, productivity and species composition.

The competition between ants and rodents appeared to be very asymmetric: the removal of rodents led to a 71% increase in the number of ant colonies, whereas the removal of the ants led to only a 20% increase in the number of rodents. This illustrates two problems in such field experiments, however: exactly how does one compare two species which differ in both size and initial abundance? We first have to decide upon a scale of measurement. Should we choose number of individuals? Number of colonies? Biomass of individuals? Table 2.2 compares the number of ant colonies with the number of rodents – a very practical measure to use, but one that implies that an ant colony is the equivalent of a rodent – a proposition that only a few would be willing to entertain (Hofstadter and Dennett, 1981). Brown and Davidson (1977) do show that if biomass is used instead of number to measure the response of rodents, then the treatments increase to nearly 30% rather than the 20% in Table 2.2.

Using an identical scale for both species to interpret asymmetry could still be misleading. Apparent asymmetry could be the result of rodents being more abundant than ants, thereby leading to a greater effect when they are removed. Perhaps rodents and ants affect each other symmetrically if the effect is expressed per gram of rodent or ant biomass. The measurement of asymmetric competition is a problem which is explored in more detail in Chapter 6.

A third taxonomic group was also involved in this interaction: birds. Unfortunately, the experiment was not initially designed to examine this factor, but observations indicated that birds foraged more on plots where both ants and rodents had been excluded, which, as Fig. 2.9 shows, were the plots where seeds had accumulated.

Lastly, one might ask how competition among rodents themselves fits into this picture. In the later series of studies in the Chihauhuan Desert, semipermeable fences were used to explore the interaction between five small granivorous rodents and three larger species of kangaroo rats. The removal of the kangaroo rats resulted in the smaller species more than doubling their combined population density. On a species-by-species basis all five smaller species increased in abundance, and four of these increases were statistically significant (Brown *et al.*, 1986).

Galindo (1986) has criticized both the experimental design and data analysis in the above studies, and concluded that they do not show any density response in rodents to the removal of ants. He notes that there were only two replicate plots, and in each the effects of ant removal were insignificant. Also, there are no pretreatment data, so we have no way of knowing whether the patterns observed were the result of differences existing before the treatments were applied. In their reply, Brown and Davidson (1986) concede that 'any rigorous reanalysis of the data probably will not sustain with a high level of statistical confidence our conclusion that rodents increased on plots from which ants had been removed'. They also point out that since the effect is asymmetric, the impact of rodents on ants is greater than the effects of ants on rodents, so the latter effect is more difficult to detect. Such controversy regarding a highly cited study illustrates the care which must be taken in both designing and interpreting field experiments.

2.3.4 A gradient of competition intensity on a shoreline

In nature, individuals often compete with a constellation of species in various combinations and densities. Both the kind and the density varies from one site to the next. We can either regard these as nearly insurmountable obstacles to predicting the intensity of competition in nature, or we can ask whether these very observations present us with opportunities for new exprimental designs. In the following example two questions are asked. Is it possible to measure the competition intensity in a plant community without resorting to pairwise removals? If it is, does competition intensity vary in a systematic or predictable way?

The system studied was a standing crop gradient on a lakeshore. Standing crop gradients on lakeshores are created by waves. Waves create a gradient in soil fertility running from infertile, sandy, exposed shores to fertile bays where silt and clay is deposited. At the same time direct disturbance to plants is highest on the sandy shores and least in the bays. Thus, the standing crop increases from disturbed infertile areas to undisturbed fertile areas. These shorelines are seasonally flooded, being submerged in May and June after the spring snow melt, but emerging as water levels fall during the summer. Most of the plants have buried rhizomes for overwintering, rapid growth during low water, and dormancy from October to May. They include an array of sedges; grasses and forbs. On the least fertile sites, evergreen rosette species and carnivorous species occur (Keddy, 1983).

To measure the intensity of competition at any point along this gradient, S. D. Wilson and Keddy (1986a) used pairs of plots, one cleared and the other uncleared. In each plot they used transplanted individuals as a bioassay of the effects of competition ('phytometers' *sensu* Weaver and Clements, 1929; Clements, 1935).

Early in the spring, 30 test individuals were transplanted into each cleared plot and each control plot. After one growing season these 30 individuals were removed and weighed. The null hypothesis is clear; if the existing plant community did not reduce performance of the transplants, then the mean performance (as measured by accumulated biomass) should be the same in the two plots. If the transplanted individuals in the vegetated plot performed less well than in the cleared plot, there was evidence of competition, and the greater this difference is, the greater the intensity of competition.

This experiment was carried out at eight positions along a standing crop gradient, and Wilson and Keddy found that the intensity of competition on this shoreline increased with standing crop (Fig. 2.10, top). Since standing crop was predictable from soil organic matter content, competition intensity could also be predicted from measurements on the soil (Fig. 2.10, bottom).

The strengths of this approach are that it completely bypasses the pairwise approach and is independent of detailed measurements of species composition in the plant community. It also exploits, rather than ignores, the fact that plant communities vary spatially. Many ecological models predict these sorts of patterns in competition intensity (for example, Connell, 1978; Grime, 1979; Huston, 1979), but this was the first field test where competition gradients were measured.

Other studies may be able to use transplanted individuals to measure competition intensity. Marked individuals could be introduced in a wide range of circumstances, and their performance measured relative to appropriate controls. The marked individuals could be recaptured for measure without doing further damage to the populations being studied.

A weakness of this approach is that it does not consider the species

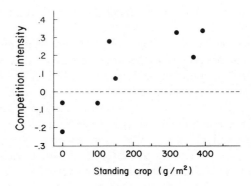

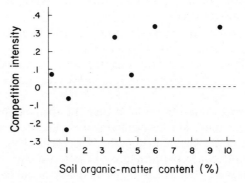

Figure 2.10 Variation in competition intensity along a shoreline standing crop gradient. The total effects of all neighbours upon the performance of three transplanted test species (*Jurcus pelocarpus, Rhynchospora fusca* and *Lysimachia terrestris*) is plotted against standing crop (top) and soil organic matter content (bottom). (After S. D. Wilson and Keddy, 1986a.)

composition of the vegetated plots, combining all species and individuals as standing crop. If species composition (rather than biomass alone) were more important, information could be lost. The relative value of species composition and biomass values in predictive plant ecology remains to be explored.

Wilson and Keddy also assumed that because of the large number of species on lakeshores, they were measuring diffuse competition. The inference about 'diffuse' as opposed to 'predominant' competition could have been avoided by using the term 'competition intensity' to describe their results.

2.4 LITERATURE REVIEW OF CASE STUDIES

Two recent review papers summarize the results of such experiments. Schoener (1983) reviewed 164 published studies of competition that were based upon field experiments, and Connell (1983) reviewed 72 such studies. Both found abundant evidence of interspecific competition – over half of the

Table 2.3 The frequency with which competition was found to be 'always present'/'always absent' in 164 field experiments of competition (modified from Schoener, 1983, Table 3). Note the uneven distribution of studies among systems and trophic levels

| | Habitat | | |
Feeding group	Freshwater	Marine	Terrestrial
Producers	2/1	13/2	74/22
Herbivores	1/6	18/6	26/9
Omnivores	0/1	–	15/7
Filter feeders	7/2	16/15	–
Carnivores	14/3	5/2	21/14
Scavengers, detritivores	0/0	1/1	7/0
Deposit feeders	0/0	2/0	–

species studied showed clear evidence of it. Table 2.3 categorizes the papers that Schoener reviewed by trophic type and ecological system. Some groups (terrestrial producers, herbivores and carnivores) have been heavily studied, whereas others (freshwater producers, herbivores and omnivores) have been little studied. Schoener also tested whether the importance of competition varied among trophic levels. The prediction, based upon Hairston *et al.* (1960) was that carnivores, decomposers and producers should compete, whereas herbivores should have their abundance controlled by predators. This hypothesis was supported for terrestrial and freshwater systems, but not for marine systems.

The results of such reviews raise methodological questions. Does a sample of published papers tell us anything about the importance of competition in nature? It is an elementary statistical point that biased samples do not provide accurate inferences about target populations. Table 2.3 shows that there is no systematic or representative selection of habitats and species on which to base generalizations. Schoener (1983) and Connell (1983) recognize other potential problems in this approach as well. If ecologists suspect that competition is not occurring in particular circumstances, then they may not even start an experiment to test for its presence. If an experiment yields negative results, the researchers may not bother to write it up and, if they do, the editor may not accept it. Similarly, in cases where pairs of organisms interact asymmetrically, there is always the temptation to manipulate only the dominant, so that the negative effects on the subordinate will be demonstrated. No amount of care in a library can overcome the fundamental problem that the literature itself, the population of studies being sampled, is itself not representative of nature. We have no way of knowing how much this problem distorts the conclusions of the above papers.

Ferson *et al.* (1986) have carefully compared the results of Schoener and Connell to try to resolve certain discrepancies between the reviews. For example, although both authors had access to the same literature, Schoener found that 77% of the species studied experienced competition, whereas Connell found that only 55% did so. Ferson *et al.* investigated two possible causes. They concluded that interpretational differences accounted for little of the discrepancy, and attributed most of it to 'methodological differences in data selection of primary research papers'. They conclude that statistical comparisons of published papers are not the final word in settling debates about interspecific competition.

2.5 THE CURRENT SITUATION

This chapter started with two possible approaches to finding general principles about competition. The first, beginning with thermodynamic principles, provides broad generality, but may provide few specific predictions about real ecosystems and whole organisms. The other approach, to work haphazardly, selecting species, habitats and experimental factors according to convenience and natural historical preferences, provides detail but little generality. With a few notable exceptions, a majority of studies appear to fall into this latter category (Connell, 1983; Schoener, 1983). After patiently accumulating more such evidence, it might be anticipated that a future scientist will assemble the results of these studies and find general patterns.

The haphazard collection of examples leading to inductive conclusions about patterns in nature can be criticized on two grounds, the first philosophical and the second very practical. First, there are fundamental philosophical problems with the process of induction (for example, Aune, 1970; Magee, 1973). Secondly, the process is inefficient and distracts us from beginning the search for general principles right now. This is an immediate concern. This is a critical period of human history, with increasing extinction rates, the rapid loss of biomes such as tropical rainforest and a growing number of international environmental problems (for example, Ehrlich and Ehrlich, 1981; Myers, 1985). If our discipline is to offer any advice on such problems, then we might reasonably plan our research to do it now rather than wait patiently for another century.

We might impatiently demand a research strategy which provides generality and general principles, but obtains them from field studies of actual ecosystems. However, before planning new research it is essential to understand clearly what we now know, and how we have learned it. The evidence presented in this chapter shows that we have learned an important general conclusion about competition: it occurs in a wide variety of species and habitats, but is not always present. We can provisionally assign ourselves the task of developing models to predict when and where competition will occur, and how intense it will be when it does occur.

The next logical step is therefore to examine some existing models, their objectives and how they have shaped the conceptual approaches used by ecologists. Chapter 4 then explores existing research strategies for testing and refining such models.

2.6 QUESTIONS FOR DISCUSSION

1. Do concepts borrowed from the thermodynamics of molecular assemblages apply to communities of organisms?

2. What do the reviews by Connell and Schoener tell us about the kinds of organisms and habitats favoured by ecologists?

3. Is there another example that you would have included if you had written this chapter? Justify its importance. (If you feel strongly about it, write to the author.)

4. 'Removal experiments demonstrate density-dependence, but density-dependence does not demonstrate competition'. Do you agree or disagree? Explain.

5. How would you test the author's assertion that ecologists have been biased when choosing the species used in their experiments? Could you do a better test than that offered in Chapter 8?

6. Which species and habitats require additional experimental study?

7. Should ecologists place priority on the collection of facts or the development of theory? Read Magee (1973), Kuhn (1970) and R. H. Peters (1980a, b). Does this change the way in which you answered the question?

3 Modelling competition

... he was very sorry to part with his sheep, which he
left with the Academy of Sciences of Bordeaux, which
offered as the subject of the prize of that year the
question why the wool of this sheep was red; and the
prize was awarded to a scholar from the North, who
demonstrated by A plus B, minus C, divided by Z, that
the sheep must be red and die of scab.

Voltaire (1759)

To construct appropriate mathematical models of
ecological processes, is, of course, only one half of
an ecologist's labours. No less important is their
testing, but the matter seems to get much less than
half of most workers' attention.

E. C. Pielou (1972)

Of course there are good models of the world and bad
ones, and even the good ones are only approximations.

R. Dawkins (1976)

All scientific knowledge can be thought of as a model of reality, a model which
is continually updated as new information accumulates. Generally the word
model is used in a narrower sense to mean a mathematical or graphical
description of some aspect of nature. There is a great deal of confusion
surrounding the value and application of models, largely because models are
constructed for different reasons, and a model constructed for one purpose
cannot always be used for another.

This chapter explores some of the models that have been used to explore
competition in populations and communities. These models are important not
so much because of their mathematical rigour, but because models can shape
the way in which questions are asked and determine the kind of experiments
performed and data collected. The principal objective of this chapter is not to
survey all the mathematical models of competition. Rather, the objective is to
provide an overview of the role of models in the study of competition, and a

perspective on the way in which these models have influenced experimental and descriptive studies of the real world. The practical examples of Chapter 2 and the theoretical framework presented here will provide the foundation for the remainder of the book.

Models can have positive roles in ecology by forcing us to state our assumptions clearly and by assisting us in exploring the logical outcomes of these assumptions. These and other virtues of models are explored below. However, models also have costs. Simberloff (1983a) writes of the 'bloated theoretical literature' when complaining that many models contribute little to the development of ecological theory. (In fairness, one could equally conclude, as I do in Chapter 4, that simply collecting another descriptive data set may be equally irrelevant to the long-term goals of ecology.) There can also be a tendency to assume that because mathematical models are strictly logical, or because they are explored with huge computers, that their conclusions correctly reflect biological reality. The basic thesis of this chapter is that models are tools. What we need to do is to sort them. There are at least three possible categories. Some may be demonstrably valuable and should be used and refined. Others are perhaps being kept only for sentimental value. Still others, new and untested, may seem to be sufficiently important that it is worthwhile testing their assumptions or predictions; given that it is often easier to spawn models than to test them, care will have be taken about assigning models to this third category.

3.1 KINDS OF MODELS

Starfield and Bleloch (1986) provide an excellent introduction to the topic, and suggest five reasons for constructing models: (1) to define problems; (2) to organize thoughts; (3) to understand data; (4) to communicate and test understanding; and (5) to make predictions. Models constructed for one purpose may be quite unsuitable for another. We use a simpler classification here, and recognize three purposes: prediction, exploration and description. Consider these in turn.

Predictive models are designed to predict the future states of systems based on relationships specified within the model between predictor (independent) variables and the predicted (dependent) variables. The dependent and independent variables must be measurable. Mechanistic relationships are not necessarily assumed. Success or failure of the model is easy to judge: the more accurate the prediction is, the better the model (Rigler, 1982).

Exploratory models can be derived from predictive models, but often have a completely independent origin. Such models allow the logical consequences of changes in assumptions or initial conditions to be explored systematically. The distinction is that the scientist is using the model to assist in a thought experiment rather than with the goal of simple prediction.

Descriptive models are used to summarize existing knowledge about the behaviour of a system. The picture may not be sufficiently complete to permit the model to be used to make predictions, but the summary can serve as a foundation for future work, or as a guide to possible experiments.

A variety of other classifications could be used. For example, Pielou (1977) proposes that mathematical models can be classified dichotomously using four criteria for a total of 16 possible styles of models. She uses the criteria: (1) whether it treats time as continuous or discrete; (2) whether it is an analytical or a simulation model; (3) whether it is deterministic or stochastic, and (4) whether it is inductive (empirical) or deductive (theoretical). However, this classification emphasizes the actual construction of the model rather than the objective for which it is used. Thus, the simpler division into three based upon modelling objectives will be used for examining models of competition.

In designing models, an investigator is faced with many trade-offs. The more precise the model is made, the more it incorporates the details of a specific system, the greater the possibility that accurate prediction is possible. However, as the model is finely tuned to one situation, there is an inevitable loss of generality. The skill of the modeller determines the degree to which a model combines generality and accuracy. Empiricists are frequently criticized for not appreciating the importance of general models, whereas modellers are often criticized for their lack of attention to reality.

Starfield and Bleloch (1986) and Holling (1978) have addressed this by considering two axes along which models can be arranged: degree of understanding, and volume of data. When the degree of understanding is low but the volume of data is high, models may be used to search for patterns and test hypotheses about them. In contrast, the degree of understanding may be high (in the sense that there is some understanding of the structure of the problem), but there may be insufficient data with which to work. This is frequently the case with ecological problems, and in such cases ecologists and ecological models are presented with two daunting challenges. First, management decisions may have to be made despite the lack of data and understanding. How do we make good decisions under such circumstances? Secondly, how do we most efficiently collect the information necessary to improve our understanding and predictive ability? Starfield and Bleloch (1986) address both of these problems.

Let us now consider some of the specific models which have shaped research on competition. The first, the Lotka–Volterra models, have been extensively covered in introductory textbooks. They are included here for completeness, but can be skipped by those familiar with them.

3.2 THE LOTKA–VOLTERRA MODELS

The Lotka–Volterra models are so popular in the study of ecology that the study of the equations themselves is frequently recognized as ecological research (Simberloff, 1983a; Fagerstrom, 1987). They can be classified as exploratory models since they assist us in thinking about how competing organisms might change in population size as a function of time and population sizes of competitors.

In introducing the use of these equations for the study of competition, Lotka (1932) begins with the concept of exponential growth. That is, in the absence of any restraining influences, the rate of growth of a population is assumed to be proportional to existing population size:

$$dN/dt = rN \tag{3.1}$$

Since this is clearly unrealistic, we assume that limitations in resource supply set an upper limit to population size, termed carrying capacity, K:

$$dN/dt = rN[(K - N)/K] \tag{3.2}$$

Note that when population size is very small (N near zero), the population growth rate is close to that of Equation (3.1) and the population is growing exponentially. When population size is very large (N near K), then population growth rate is nearly zero.

To explore competition, two equations must be examined simultaneously. What is needed for each population is a measure of how it is affected by the presence of the other. Consider population 1. The effect of each individual of population 2 upon population 1 can be measured by comparing the *per capita* effects of population 2 relative to those of population 1. This is done with an entity called a **competition coefficient.** If individuals of population 1 have the same effect on population 1 as individuals of population 2, then the competition coefficient equals unity – that is, the two populations are indistinguishable from the point of view of population 1. However, if individuals of population 2 have much greater effects upon individuals of population 1 than individuals of population 1 have upon each other, then the competition coefficient is greater than unity. Conversely, if the per capita effects of population 2 are much less that the effects of population 1, then the competition coefficient is less than unity. By proceeding with analogous arguments for population 2, one ends up with two differential equations. The growth rate of each population is then determined both by its own population size and that of the other population, with the effects of the latter weighted by its competition coefficient (a_{ij}) as follows:

$$dN_1/dt = r_1 N_1 [(K_1 - a_{11}N_1 - a_{12}N_2)/K_1] \tag{3.3}$$

$$dN_2/dt = r_2 N_2 [(K_2 - a_{22}N_2 - a_{21}N_1)/K_2] \tag{3.4}$$

By definition, each species' competition coefficient upon itself is unity. Since there are two competing populations, two outcomes are possible when these populations are allowed to grow, interact and reach equilibrium: (1) one species becomes extinct and the other climbs to its own carrying capacity; or (2) the species coexist.

A principal objective of studying these equations has been to determine what controls whether the two species coexist, and, if they fail to do so, which of the two will be the winner. Since the only parameters to work with are the values for $r_1, r_2, N_1, N_2, K_1, K_2$ and a_{12} and a_{21}, the solution is given in terms of these.

One way to picture this two-species interaction is using species isoclines. An **isocline** is simply all possible sets of conditions where the growth rate of a population is zero. In the case of this model it is all possible pairs of population densities where the growth rate of one population is zero. The isocline is derived by setting the population growth rate equal to zero:

$$dN_1/dt = r_1 N_1 [K_1 - a_{11} N_1 - a_{12} N_2)/K_1] = 0 \qquad (3.5)$$

For this population, the growth rate is zero when one of three conditions is satisfied:

$$r_1 = 0 \qquad N_1 = 0 \qquad (3.6)$$

$$K_1 - a_{11} N_1 - a_{12} N_2 = 0$$

Obviously if r or N equals zero, the rate of population growth will be zero and thus these are trivial solutions. It is the third condition which is of interest. We can plot this isocline on a graph by solving for the intercepts of the axes and joining them with a straight line. The intercept with the N_1-axis is derived by setting N_2 equal to zero, in which case

$$K_1 - a_{11} N_1 - a_{12} 0 = 0$$

$$K_1 - a_{11} N_1 = 0$$

$$K_1 = a_{11} N_1$$

$$N_1 = K_1/a_{11} \qquad (3.7)$$

Similarly, the intercept with the N_2-axis is determined by setting N_1 equal to zero. What we are doing biologically is asking how many individuals of either population 1 (N_1) or population 2 (N_2) are required to produce a zero growth rate for population 1. These results can be plotted as in Fig. 3.1. Above this isocline population 1 has exceeded the carrying capacity, so the population size declines with time. Below the isocline the population size gradually increases. At any point along the isocline the growth rate is zero and the population size remains constant. This simple plot enables us to explore how the size of population 1 will change under all possible sets of conditions represented by different sizes of the two populations.

To explore the interaction of both populations simultaneously, we must go through the identical series of steps for population N_2, and plot its isocline in an identical manner. This is where things start to become interesting, and we can begin to talk about the behaviour of the model, for the two isoclines can be arranged in different ways. Figure 3.2 shows the four possibilities, and each of these arrangements of isoclines has different consequences for the mixture of the two species.

In the top two cases only one species survives at equilibrium; i.e. there is a

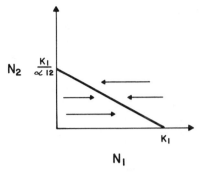

N_2 $\dfrac{K_1}{\propto 12}$

K_1

N_1

Figure 3.1   The isocline for population 1 plotted in two-dimensional space, showing all possible combinations of N_1 and N_2 (that is, population sizes for populations 1 and 2) where the population growth rate of N_1 is zero. As the arrows show, above this line population size falls to the isocline, and below the line population size increases to it. A similar isocline can be drawn for population 2.

competitive dominant and a subordinant, with the dominant being the species with the isocline furthest from the origin. We must specify at equilibrium because if we start at some arbitrary mixture of the two populations, both will co-occur until the trajectory collides with the axis of the dominant. Since ecologists frequently use these models to study coexistence under equilibrium conditions, these top two models often receive the least attention.

In the third case the two species coexist. This is because each species is more negatively affected by intraspecific competition than by interspecific competition. It is regularly assumed that intraspecific competition is indeed higher than interspecific competition, based on the assumption that more similar individuals compete more intensely, with conspecifics being most similar to each other. Thus, this outcome has a certain appeal based on its assumed biological reality. Also, since many modellers are most comfortable with the assumption that nature is at equilibrium (even if we know that this is most certainly not the case), this outcome has a certain mathematical appeal. Thus, this outcome is frequently given more attention than the above two. Questions such as 'How many species can coexist at equilibrium...?' can be explored using this model.

In the final situation, the winner of the two-species competition can be predicted only when starting population sizes are known. Yodzis (1978) calls this 'contingent' competition. The intensity of interspecific competition is such that, once a species begins to achieve numerical superiority, it damages the other so severely that the outcome becomes certain. Which population achieves this initial superiority depends solely on the assumed starting density. Under these circumstances, like the first two, we can again predict with certainty that only one of the populations will persist at equilibrium. Thus, the model is again unsatisfactory for the study of coexistence under equilibrium

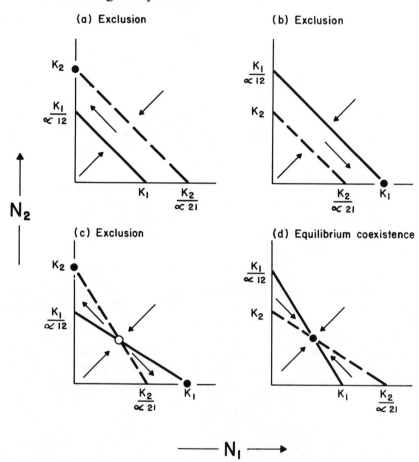

Figure 3.2 Four possible arrangements for the isoclines of two competing populations. The arrows show, as in Fig. 3.1, changes in population size with time. The solid
dots represent the equilibrium points (expected outcomes) of these pairwise interactions. Three of the four possible arrangements (a)–(c) result in competitive exclusion.
(a) Exclusion; competitive dominance by population 2. (b) Exclusion; competitive
dominance by population 1. (c) Exclusion; contingent competition (Yodzis, 1978). The
winner depends upon the initial starting densities. The open circle is an unstable
equilibrium point which is of greater mathematical than biological interest.
(d) Equilibrium coexistence; *not* to be confused with 'coexistence' which includes many
other mechanisms that prevent competitive exclusion (e.g. Fig. 7.1).

conditions. It is also unsatisfactory because it suggests that the inherent
biological characteristics of the two species do not allow us to determine their
behaviour in mixture. Note that in theory coexistence is possible, in that there
is an equilibrium point where the two isoclines intersect. This is an unstable
equilibrium point, however, in that as soon as the populations diverge from

this precise mixture of population sizes, they move inexorably towards exclusion. Thus, this equilibrium point is of limited mathematical or biological interest.

In a more rigorous manner, we may specify the above outcome in terms of the carrying capacities (K) and competition coefficients (a_{ij}) of the two populations. Since both may vary, it is easiest to picture this by assuming the carrying capacities of the two populations are identical $(K_1 = K_2)$. In cases where they require the same resources, and we use an appropriate measure of carrying capacity such as biomass, this is not unreasonable. Under such circumstances we can specify the three outcomes as follows:

Exclusion (competitive dominance) (Fig. 3.2(a) and (b))

$$a_{11}/a_{21} < 1 \quad \text{and} \quad a_{12}/a_{22} < 1 \tag{3.8}$$

$$a_{11}/a_{21} > 1 \quad \text{and} \quad a_{12}/a_{22} > 1$$

Exclusion (contingent competition) (Fig. 3.2(c))

$$a_{12}/a_{22} > 1 > a_{11}/a_{21} \tag{3.9}$$

Equilibrium coexistence (Fig. 3.2(d))

$$a_{12}/a_{22} < 1 < a_{11}/a_{21} \tag{3.10}$$

Roughgarden (1979, Fig. 21.4) illustrates for each of the three above outcomes the actual trajectories which pairs of populations will follow given different starting population sizes.

Variation in carrying capacity can be superimposed on these relationships. For those who think graphically, Vandermeer (1970) has provided an elegant demonstration of these relationships (Fig. 3.3). In this figure the isoclines are

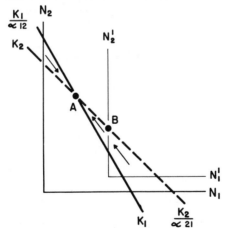

Figure 3.3 The effect of changing carrying capacity on a two-species system. Using axes N_1 and N_2, stable equilibrium occurs at point A. Using axes N_1' and N_2' population 2 excludes the other and achieves dominance at point B. (After Vandermeer, 1970.)

plotted relative to different axes. Displacing the axes is equivalent to changing the carrying capacity and, as the figure shows, a situation with a stable equilibrium point (A) can be converted to one with competitive dominance (B) as carrying capacities are changed.

3.2.1 The community matrix

The two-species situation can be expanded to n interacting species, in which case the equation for each species is expanded to include all other species. Thus, the equation for population 1 expands to

$$dN_1/dt = r_1 N_1 [(K_1 - a_{11} N_1 - a_{12} N_2 - \cdots - a_{1n} N_n)/K_1] \quad (3.11)$$

where the equation includes a competition coefficient for each of the n species with which it is possible to interact. For the case of three species, each isocline becomes an isoplane in three-dimensional space. The competitive dominant in such a three-species system would be the one with an isoplane farthest from the origin. If the planes were tipped such that they intersected, stable lines rather than stable points can be imagined. Situations with more dimensions are usually represented with a matrix of competition coefficients called a **community matrix** (Levins, 1968; Yodzis, 1978). In this matrix (Fig. 3.4) each row lists all of the competition coefficients determining the population growth rate of the species represented by that row. In the same way each column vector lists all of the impacts which that species has upon the growth rates of neighbouring populations. By considering rows or columns, one can think either in terms of the effects of all species upon a selected species of interest, or of the effects of a selected species upon all possible neighbours.

Such matrices appear to be powerful tools for comparing the different kinds

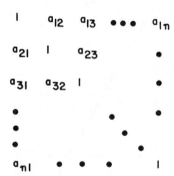

Figure 3.4 A community matrix. Each competition coefficient a_{ij} measures the effect of an individual of population j upon an individual of population i. Each row describes how individuals of other populations reduce the population growth rate of the population represented by that row. Each column describes how one population reduces the growth rates of all other populations.

of communities which exist in nature, and exploring how they may respond to perturbations. To date, much of the emphasis has been upon the mathematical stability of such systems. That is, are there stable points, and how do model communities respond when perturbed away from such stable points (Levins, 1968; May, 1974)? In this context, stability occurs when the net population growth rates of all populations, dN_i/dt, equal zero. If such a system is perturbed, it may return to the stable point, it may continue to diverge from it or it may remain at the point to which it was perturbed. If it does change with time, it may do so monotonically or else through oscillations (May, 1974). Such questions assume either that nature is sufficiently near to equilibrium to make such questions relevant, or at least that stability is a useful reference point for studying real systems.

Since many natural communities are likely to be far from equilibrium, we may ask what other roles exist for community matrices. Yodzis (1978) has provided an exploration which places much less emphasis upon stability. Instead, he asks questions about the different kinds of community matrices which can exist, the sort of biological interactions which will exist in each, and finally (although it is not considered here) how each will respond to harvesting. Yodzis begins by contrasting competition for space with the competition for other kinds of resources. He proposes that although competition for many resources may be symmetric, competition for space is likely to involve interference competition and therefore be asymmetric. He emphasizes the importance of competition for space in ecological communities, noting that space will be particularly important as a resource for sessile organisms such as corals and plants, but also for the many kinds of animals which are territorial.

Yodzis then generates model communities using competition coefficients selected randomly according to certain constraints. He explores two basic types of communities. In the first kind, the competition coefficients are all greater than unity. This means that individuals of each population damage individuals of other populations more than themselves. As a consequence the first population to arrive and colonize a site holds it against all other populations. Thus, although competition is very definitely present in such communities, the distribution and abundance of populations is a consequence of their initial colonization patterns. Yodzis therefore calls these 'founder controlled' communities.

A second type of community matrix which Yodzis explored consisted of pairs of competition coefficients in which there are many asymmetric interactions. (Given that the competition coefficients were generated at random subject to certain constraints, such matrices also probably included coexistent interactions.) In this case, although initial colonization patterns initially determine the distribution and abundance of populations, competitive dominants gradually exclude their neighbours. Yodzis calls these communities 'dominance controlled'.

Several important points emerge. First, Yodzis draws our attention to the fact that matrices with different combinations of competition coefficients have different kinds of biological behaviour. This suggests the research strategy of asking what kinds of matrices occur in nature and what the consequences might be for the organization of such communities. Secondly, he notes that this may allow us to make predictions about how these communities will respond to natural perturbations such as harvesting. Lastly, he proposes that competition for space may be fundamentally different from competition for other kinds of resources, and encourages us to think about patterns of community organization that can be shared by very different kinds of organisms.

Although such matrices can be constructed artificially for exploratory modelling, producing actual community matrices is difficult. One procedure involves using measures of niche overlap to produce competition coefficients; this is invalid, as discussed below. An alternative is to use experimentally measured values from large competition experiments. This approach is considered in more detail in Chapter 6.

3.2.2 Relationship with biological reality

The Lotka–Volterra equations should be considered exploratory models with limited direct relationship to real ecosystems, and their popularity is probably at least partly attributable to sentimental attachment. Some obvious weaknesses are the unrealistic assumptions of the model, such as those that individuals are all equivalent, making age or size class structure irrelevant, and that individuals are thoroughly mixed so that they all influence each other directly and equally. The principal problem in applying these models to predict the behaviour of real communities lies in assigning meaningful values to the coefficients. This is so difficult as to be impossible for many systems. In the case of uniform environmental conditions (constant competition coefficients), the number of coefficients to be estimated is the square of the number of populations, so a comparatively simple community with 10 species requires the estimation of 100 competition coefficients. In nature the intensity of competition may vary with a range of environmental factors including climate, kind of resources, spatial distribution of resources and temporal variation in all of the foregoing. Thus, the coefficients themselves become variables.

In addition there is a second problem in estimating competition coefficients. Each competition coefficient for a row in a matrix is scaled relative to intraspecific competition for that species (for example, Begen and Mortimer, 1981). Intraspecific competition is assumed to be equal to unity for each species; i.e. in the community matrix it is assumed that the diagonal matrix consists of ones. However, with communities made up of very different species, there is no obvious reason why intraspecific competition should be the same for them all. In such cases we might expect intraspecific competition to be much more intense in some species than in others. If we then attempt to add up the effects of a species down a column vector, we are comparing competition coefficients each measured on a different scale. For a simple two-species system of very similar species, this assumption may not be far from biological reality, but the more we try to work with entire communities, the more biologically unrealistic the assumption seems.

These models can also misdirect research. For example, ecologists have been driven to understand the factors which determine the number of species which can coexist in a given area (May, 1986). The Lotka–Volterra equations direct attention to the points of stable coexistence. However, there is an important distinction between coexistence and stable coexistence. In a world where the environment constantly fluctuates, non-equilibrium coexistence is more biologically plausible (for example, Huston, 1979; Grime, 1979; Pickett and White, 1985) even if it is less mathematically tidy. Yet recent reviews on competition such as Arthur (1987) still deal largely with stable coexistence. This illustrates the power that models can have in determining the sorts of questions that ecologists consider interesting.

3.3 A RESOURCE COMPETITION MODEL

Tilman (1982) has introduced an alternative model for competition. This model emphasizes the mechanisms of species interactions, which he contrasts with the more 'phenomenological and descriptive approach of classical theory'. In the Lotka–Volterra model, the behaviour of two species was described using six constants: the inherent growth rates, carrying capacities and competition coefficients for each species. Six different constants are needed to explore the resource competition model proposed by Tilman.

Only one of his models is considered here: the case of two species competing for two essential resources. Recall that essential resources were defined as non-substitutable resources in Chapter 1. A good example would be the requirement of plants for both light and nutrients, where one cannot be substituted for the other.

We begin by considering the growth rate of one species in two-dimensional space. However, in this case the two dimensions refer to the relative abundances of the two resources rather than to the abundance of the two species. As with the Lotka–Volterra model, we explore the behaviour of a single species by deriving its isocline, and then move to superimposing the isoclines of two species. A species' isocline is determined by specifying all possible sets of resource levels which produce zero net growth. In the case of essential resources, the isocline will look like that in Fig. 3.5. Below a critical level of either resource the population size declines. In this case the critical levels are marked as R'_1 and R'_2. For all possible resource levels in the hatched area, growth is positive. If either resource is at the critical minimum level, growth is halted – thus the abrupt right angle in the isocline.

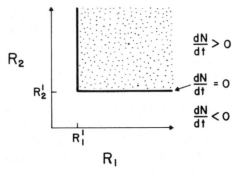

$$\frac{dN}{dt} > 0$$

$$\frac{dN}{dt} = 0$$

$$\frac{dN}{dt} < 0$$

Figure 3.5 The response of a population to two essential resources, R_1 and R_2 (recall Fig. 1.3). The thick line marks the zero net growth isocline where resource levels (R'_1 and R'_2) are just sufficient to maintain the existing population size. Above the critical levels R'_1 and R'_2 the population will grow (stippled area); below these resource levels the population size declines.

Once the minimum necessary conditions for growth are specified, the rate of resource supply (or renewal) is considered. The biological argument here is that all habitats have a rate of resource renewal. In the case of plants it would be the rates of addition of elements like phosphorus through weathering and rainfall. In the case of scavenger beetles it would be the rate of death in small mammals. In the case of dung beetles it would be the rate of defecation in large mammals. In desert rodents it would be the rates of seed production by plants. In any habitat it should be possible to measure the supply rates of key resources. In a two-resource system we need to specify a supply vector, **u**, which specifies the rate of renewal for R_1 and R_2. This resource supply vector for the habitat is illustrated in Fig. 3.6 (left). The resource supply vector may vary with position along the two resource axes (Fig. 3.6, right). Obviously the

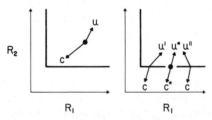

Figure 3.6 (left) The contrasting effects of resource supply rate (**u**) and resource consumption rate (**c**) on the resource levels in a habitat (R'_1, R'_2) indicated by the solid dot. The population can be pictured pulling the characteristics of the habitat (solid dot) towards the lower left by converting resources to biomass, whereas the supply rates pull it towards the upper right. The zero net growth isocline is also shown; in this case the population will continue to grow until the consumption vector (**c**) pulls the solid dot to the zero net growth isocline. (right) The equilibrium point occurs when the environment (solid dot) has been pulled to the zero net growth isocline and the resource supply vector exactly balances the resource consumption vector (**u***, **c***).

resource supply vector, assuming it is positive, will gradually increase the resource levels above the isocline. Once this occurs the population will being to grow.

Once the population begins to grow, it begins to consume resources. This leads to the final constant which needs to be considered: the resource consumption vector, c. If a species is consuming two essential resources, one could measure the amount of resource consumed per unit time. A simple way of doing this might be simply to analyse the organism for the level of the two resources in its tissues. Alternatively, one might actually try to monitor the rates of resource depletion by determining rates at which dung beetles degrade dung or rodents consume seeds. If this is done, the result is the vector c called the resource consumption vector. Typical resource consumption vectors are illustrated in Fig. 3.6.

It should now be possible to picture intuitively the behaviour of this model. The current state of the environment is specified by the dot in Fig. 3.6 (left), and the resource consumption vector and resource supply vectors engage in a tug-of-war, pulling the environment around the two-dimensional space. If the rate of consumption exceeds the rate of supply, then the resource levels will gradually decline (i.e. the state of the environment will drift from upper right to lower left) until the isocline is intersected. At this point the growth of the population stops, halted by whichever resource is limiting at that isocline.

Where will it all end? It is obvious that equilibrium is possible if there is a point (or points) where the supply and consumption vectors are equal and opposite, i.e.

$$\mathbf{u} + \mathbf{c} = \mathbf{0} \tag{3.12}$$

This equilibrium point must be on the isocline. Depending upon the size and direction of the supply and consumption vectors, it may also be a stable point, as illustrated in Fig. 3.6 (right, middle pair of vectors).

With these constants one can model the behaviour of a single population in this two-dimensional resource space. The next step is to pose questions about the behaviour of two species sharing this space. This is done by repeating the above steps for a second species. Note that although it is necessary to derive a new isocline and new consumption vector, the supply vector will remain the same, so that only four additional constants need to be specified for the second species. The possible outcomes are shown in Fig. 3.7. They are identical to the four possibilities identified with the Lotka–Volterra model: exclusion (competitive dominance by one of the species), unstable equilibrium (contingent competition) or equilibrium coexistence.

Consider these possibilities in turn. In the first case the isocline of species A is always inside that of species B. This means that species A requires less of both essential resources. At equilibrium it will have 'pulled' the resource state to a level outside the isocline of species B, and thus species B will decline to extinction. This is analogous to the situation in Fig. 3.2(a). A similar situation

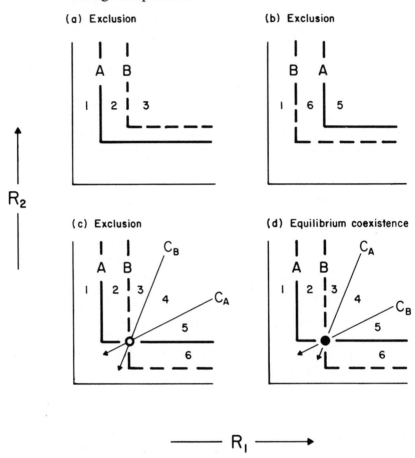

Figure 3.7 The four possible outcomes of populations A and B interacting for resources R_1 and R_2 with consumption vectors c_A and c_B. (a) Exclusion; competitive dominance by A. (b) Exclusion; competitive dominance by B. (c) Exclusion; contingent competition in region 4 (regions 1–6 are discussed in the text). (d) Equilibrium coexistence in region 4 (regions 1–6 are discussed in the text).

occurs in the second case, except that species B is dominant. In both of these cases, there is a shared requirement for the resources, and the dominant is the species that is capable of lowering resource levels to the point where the other is unable to survive.

The remaining two cases have isoclines which cross. Obviously, crossed isoclines produce an equilibrium point. In Fig. 3.7(c) the point is unstable, and in Fig. 3.7(d) it is stable. Up to this point the outcomes are little different from those of the Lotka–Volterra model. Where the behaviour deviates from the Lotka–Volterra model is the greater range of possibilities in the last two situations. In order to explore this, however, resource supply processes require more attention. For simplicity, consider the case of stable equilibrium

(Fig. 3.7(d)). We must define another point, called the resource supply point, which specifies the maximum possible levels of the two resources. It has co-ordinates (S_1, S_2). Further assume that the rate of supply of a resource is dependent upon the distance which a point (R_1, R_2) is from it. That is, assume that rate of supply is proportional to $S_i - R_i$. If this is the case, then, for any arbitrary point (R_1, R_2) the resource supply vectors will point towards (S_1, S_2). Moreover, the length of the vector will vary with the distance from that point. The position of this resource supply point is then used to explore the coexistence models further. Now return to the numbered regions in Fig. 3.7(c) and (d). If the resource supply point is in region 1, then neither species can survive. If it is region 2, then only species A is capable of surviving and, in a symmetrical way, if it is in region 6, then only B is capable of surviving. Under these circumstances biological interactions are unimportant – the resource supply levels are simply outside the tolerance limits of one of the two species. When by themselves, either species can survive in regions 3, 4 and 5. When mixed, however, species A wins in region 3 and species B wins in region 5. In the case of situation (d), stable coexistence, the species will coexist if the resource supply point is in region 4. This can be understood by picturing the trajectory that the species will follow in each of these regions. In regions 3 and 5 an isocline is intersected, whereas in regions 4 they move towards the stable point. Exactly the same behaviour occurs in Fig. 3.7(c), except that in region 4 either species A or species B will win depending upon the initial conditions. In this case the equilibrium point is unstable, so that if there is the slightest departure from it one of the species will be driven to extinction; it is therefore only of mathematical interest. This is identical with the situation in Fig. 3.2(c). However, in the resource consumption model the critical factor which determines whether the equilibrium point is stable is the consumption vectors of the two species. If each species consumes relatively more of the resource that limits its own growth at equilibrium, then the point will be stable. If each species consumes more of the resource which limits the other's growth, then the point is unstable. This is exactly what occurred with the Lotka–Volterra model – there is stable coexistence only if intraspecific competition is greater than interspecific competition.

Tilman (1982) uses this model to explore questions of coexistence in plant communities, and readers are referred to his monograph for more details. The strength of this model is its emphasis upon mechanism. The picture of plants pulling the environment in different directions like stretching a piece of sheet rubber is a vivid one. A problem is that the amount of information needed to construct a predictive model is still excessive. Tilman (1982) writes

> To test this theory thoroughly, it will be necessary to know the resource requirements and competitive interactions of the dominant species under controlled conditions, the correlations between the distributions of these species in the field and the distributions of limiting resources, and the effects of various enrichments on the species composition of natural communities.

Since the model is relatively new, the verdict is not yet in on its utility. However, except for certain specific sets of conditions, it too may be exploratory rather than predictive. Applications to freshwater phytoplankton are discussed in Sell *et al.* (1984) and Tilman *et al.* (1984). Tilman (1988) has since presented a more elaborate model which considers plant responses to different resource ratios by including allocation to foraging. The inclusion of within-plant allocation is another attempt to include more mechanistic elements in competition models.

3.4 COMPETITION, BEHAVIOUR AND HABITAT USE

The above two models described the effects of competition on population sizes and resource levels. Another possible dependent variable would be the behaviour of the competing species. We know from the many examples of resource partitioning that different species use different resources, but there are many unresolved issues in the study of trade-offs in foraging (Pyke, 1984), and it is not at all clear how patterns of resource use might change under different intensities of inter- and intraspecific competition. Pimm and Rosenzweig (1981) and Rosenzweig (1981) have presented such a model.

Envisage the following situation. There are two species which occupy a region which has two resource patches (or two resources). Each species 'prefers' one patch type – that is, each is specialized to exploit one patch type more efficiently than another. However, each species can exploit both patch types when population densities are low. The model explores how the foraging behaviour of one species responds to all possible population sizes of the two species. Under which conditions will a species use both habitats, and under which conditions will it use only the one upon which it is specialized? Under which conditions will it be a generalist, and under which will it be a specialist? (The words generalist and specialist are being used in a narrowly defined sense to describe variation in behaviour or variation in realized niche width, and *not* variation in fundamental niche width.)

Consider the habitat from the point of view of species 1 in the absence of species 2. In Fig. 3.8 this is represented by the horizontal axis. If the population size of species 1 is small, then it will clearly be advantageous for species 1 to forage in the habitat upon which it is specialized. Now allow population density to grow by slowly moving to the right along this axis. As this happens, population densities in the preferred patch are increasing, as is intraspecific competition. Eventually a point is reached where intraspecific competition in the preferred patch is so intense that resources are depleted to the point where the two patches become equally preferable. If population size increases any more, the unoccupied patch type offers a better return per unit of foraging effort. This point on the axis is then marked with a dot. To the left, species 1 forages in one patch, to the right it forages in both. Now imagine that the other patch is occupied with a small number of

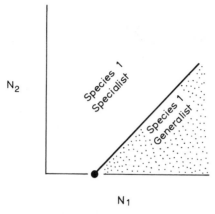

Figure 3.8   The changes in foraging behaviour of species 1 as a function of all possible population sizes of species 1 and 2. In the absence of species 2 ($N_2 = 0$, consider only the horizontal axis) and at low levels of species 1 (left-hand end of axis), species 1 is a specialist upon its preferred resource. As the population size increases (moving right along the horizontal axis), intraspecific competition for the preferred resource becomes more intense. At some point (dot) individuals begin foraging for a less preferred resource in order to avoid intense intraspecific competition; at this point the species becomes a generalist. If we allow the population size of species 2 to increase (moving up the page), the point at which individuals of species 1 expand diet or habitat to include the less preferred resource also changes, since there is now interspecific competition for this resource. Higher levels of intraspecific competition are necessary to induce the switch in foraging from specialist to generalist. This logic marks out two regions for a species – a region where it is a specialist, and a region in which it becomes a generalist (stippled).

individuals of species 2. This means that there is interspecific competition which species 1 encounters when foraging in that patch, which renders that patch even less suitable. That is, because it is already occupied by another species, its apparent quality to species 1 has declined. Now consider again how the behaviour of species 1 will change in response to increased size of its own population. Clearly the point at which intraspecific competition makes the other patch type attractive must be higher, since the other patch type is now less desirable. Thus, species 1 does not begin foraging in both habitats until its population size is somewhat higher. Following such logic, one can construct a line with positive slope which represents where the two behaviours produce equivalent returns. A decrease in the population size of species 1 favours specialist foraging: a slight increase favours generalist foraging. By analogy with the idea of population size isoclines, this line is referred to as an **isoleg**. Figure 3.8 shows the resulting behaviour of species 1 under all densities of the two species. The isoleg of species 2 can be derived in exactly the same manner.

As with the examples above, the interesting results for two-species

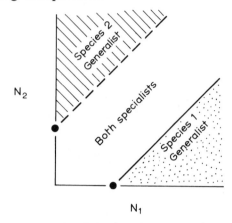

Figure 3.9 One possible foraging behaviour for two species as a function of their population densities. This is obtained by superimposing the results of Fig. 3.8 for two species. At low population densities both species forage in their own preferred patches, but as the population sizes increase, intraspecific competition forces each to forage in the other's patch. Species 1 is therefore a generalist in the stippled area ($N_1 \gg N_2$) and species 2 is a generalist in the hatched area ($N_2 \gg N_1$).

interactions are derived by superimposing the two graphs. Figure 3.9 shows the situation for two species with non-intersecting isolegs. Three regions can be recognized, corresponding to three of the four possible combinations of behaviours.

1. Both species have low population sizes and each therefore behaves as a specialist;
2. Species 1 is a generalist due to high intraspecific competition, but species 2 is still a specialist;
3. Species 2 is a generalist due to high intraspecific competition, but species 1 is still a specialist.

Pimm and Rosenzweig (1981) explore the four possible combinations of isolegs in two-dimensional space. The next interesting step is to superimpose these isoleg plots upon plots of isoclines, and Rosenzweig (1981) provides an introduction to this procedure.

It is therefore possible to explore how the behaviour of two species ought to change in response to varying degrees of inter- and intraspecific competition. This illustrates the difficulty of measuring competition coefficients from measures of realized niche overlap, because depending where one measures in Fig. 3.9, one could find no niche overlap or high niche overlap.

The model is primarily exploratory, but could be made predictive for pairs of species if it were important to predict their foraging behaviour. This would require simultaneous measures of both the range of resources consumed and the population sizes of both species. The isoleg would be mapped by determining the region where behaviour shifts abruptly. Whether it is worth this much effort to predict the foraging behaviour of two species is open to discussion. It is not clear to what degree one could extrapolate from one pair of species to another, in which case this approach becomes a complicated method of describing the interactions of pairs of species on a case-by-case basis. We must ask at some point what the priority dependent variables (or state variables) for community ecology are, a theme which is returned to in later chapters. Perhaps the foraging behaviour of two species would not qualify for a high priority.

Rosenzweig (1981) summarizes the model's testable predictions. He concludes with some general observations on the testability of quantitative ecological models.

> There is so much noise in ecosystems that it is always possible to wonder if the measurement failed to fit the theory because of the noise and not because of the inadequacy of the theory. On the other hand, theories which make qualitative predictions are often too easily fit. Most qualitative predictions simply state the existence and direction of a pattern.... All too often, even these are discovered before the theory is advanced and no further predictions from the theory are made to allow the pattern to be tested.

3.5 TWO GRAPHICAL MODELS FOR RESOURCE PARTITIONING

It is frequently assumed that all models require equations, but some useful models are graphical. Of course, these models can often be formulated mathematically, but the form in which they are generally used is pictorial. Often such models are used to summarize existing understanding of ecological processes, and to make qualitative predictions about patterns in nature. The best example of this is the picture of resource partitioning presented in Fig. 3.10, some of the mathematics of which are described in MacArthur (1972) and May (1974). The following section explores some of the impacts this picture has had upon the sorts of questions that ecologists ask and the kinds of data that they collect. The patterns of the model are first described, and then two competing models proposing to account for these patterns are presented.

3.5.1 Patterns

The pictorial representation of resource partitioning illustrates how seven species of organisms can coexist by using different sections of a resource continuum (Fig. 3.10). Each species is assumed to have a bell-shaped resource utilization curve, and the zones of overlap are assumed to show areas where there is interspecific competition for resources (for example, MacArthur, 1972; Vandermeer, 1972; Cody, 1974; Whittaker and Levin, 1975; May, 1981;

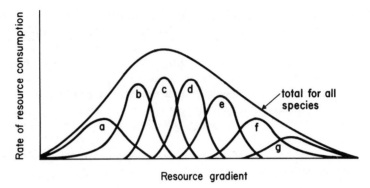

Figure 3.10   Resource partitioning in a hypothetical community comprised of seven populations (after Pianka, 1981). Each curve represents the resource utilization by a different species. The wider the curve is, the less specialized the *realized* niche of the species. Adding together the demands of all species produces the upper curve.

Pianka, 1973, 1981; Giller, 1984; Arthur, 1987). The model is so widely used in studies of competition that it deserves closer attention. In considering this model it is essential to distinguish between realized and fundamental niches. Fundamental niches are resource-use patterns which occur in the absence of competitive effects produced by neighbours of different species. They can only be measured by experimental removals of neighbours, or possibly by finding natural situations where neighbours are absent. Realized niches are resource-use patterns documented in the presence of conspecific neighbours, and if competition affects realized niche width, then realized niches will be narrower than fundamental niches. Field studies which describe species distributions and resource consumption provide information only on realized niches. Some of the confusion in the current literature arises from using words like niche without specifying whether it is realized or fundamental (for example, Cody, 1974; Pianka, 1981). A similar pair of terms, physiological response and ecological response, have been used by plant ecologists (Mueller-Dombois and Ellenberg, 1974).

The differentiation of realized niches is a basic truth of natural history. Figure 3.11, for example, shows the main dimensions of resource partitioning for African rain forest squirrels. This recent example and the many cited in Schoener (1974) illustrate the view of Hutchinson (1959) that

> the process of natural selection... leads to the evolution of sympatric species which at equilibrium occupy distinct niches.... The empirical reasons for adopting this view and the correlative view that the boundaries of realized niches are set by competition are mainly indirect.

If species coexist by using different niches, then knowledge of the number and kind of niches in a community clearly permits predictions about the number of species likely to be found there. Unfortunately, niches are usually

	HABITAT TYPE	VEGETATION HEIGHT	BODY SIZE	FOOD TYPE	ACTIVE PERIOD
Myosciurus pumilio	mature and disturbed forest	arboreal	tiny	bark scrapings	full day
Aethosciurus poensis	mature and disturbed forest	arboreal	small	some diverse arthropods	full day
Heliosciurus rufobrachium	mature and disturbed forest	arboreal	medium	some diverse arthropods	part day
Protexerus stangeri	mature and disturbed forest	arboreal	large	hard nuts no arth.	part day
Funisciurus lemniscatus	mature and disturbed forest	ground foraging	small	many termites	full day
Funisciurus pyrrhopus	mature and disturbed forest	ground foraging	medium	many termites	part day
Epixerus ebii	mature and disturbed forest	ground foraging	large	hard nuts few arth.	part day
Funisciurus isabella	dense growth	lower levels	small	leaves diverse arth.	full day
Funisciurus anerythrus	flooded forest	all levels	medium	many ants	full day

Figure 3.11 Resource partitioning by rain forest squirrles in Gabon. Horizontal lines indicate nearly complete separation between the characteristics. Food types include only those which differentiate species. Squirrels may either forage for most of the day (full day) or return to their nests several hours before sunset (part day). (After Emmons, 1980; consult Schoener, 1974, and Giller, 1984, for many other examples.)

recognizable only when species are filling them. This would not prevent important advances in predictive ecology, however, for if one can describe niches occupied in one example of an ecosystem, then one can reasonably predict that similar niches occur in other examples of that system. Thus, a periodic table of niches (*sensu* Pianka, 1983) may be possible for each community type in the biosphere. Moreover, if only a few axes are necessary to account for the niche differentiation, then general predictions may be possible. Figure 3.11 suggests, for example, that vegetation height and food type would be two main axes we would need in order to understand and predict species composition of squirrels in tropical rain forest. Similarly, Diamond (1975) proposes that branch size and fruit size will enable us to understand the composition of tropical fruit pigeon communities. Schoener (1974) and Giller (1984) provide other examples. Thus, studies of niche differentiation provide both a description of community organization and the possibility of predicting composition in other related community types.

There is the problem that there is no obvious upper limit to the number of species which can fit into a community, since one can always postulate additional niches. Consider regeneration niches in plant communities. Grubb's (1977) review certainly suggests that there are more than enough regeneration niches to account for the diversity of plants and plant communities. There are so many possibilities for niche differentiation that one is left wondering whether general theories are possible, or whether plant ecologists will become no more than natural historians painstakingly documenting the regeneration niches of each plant species in each community. Another example of this is the remarkable number of niches which insects can find to exploit on a single plant species, bracken (*Pteridium aquilinum*) (Lawton, 1984). This raises the question of whether such niches can even be recognized unless they are already occupied, and whether some communities can be predicted to have 'vacant niches' (Price, 1984b; Herbold and Moyle, 1986).

Arthur (1987) proposes testing whether resource partitioning actually permits coexistence. First, it is important to realize that resource partitioning is proposed to account only for stable coexistence; it is unnecessary for non-equilibrium coexistence. Secondly, there are five potential causes of stable coexistence, only one of which is resource partitioning (Arthur, 1987). Given this context, Arthur proposes three steps to demonstrate conclusively that resource partitioning is the cause of stable coexistence: (1) demonstration of stable coexistence in a system with a given level of resource heterogeneity; (2) demonstration of competitive exclusion in a less heterogeneous system; and (3) quantification of resource utilization functions for the two species showing that significant separation is possible in (1) but not in (2).

Resource partitioning describes an obvious pattern in nature, but how is it related to competition? Realized niches are often assumed to be different among species and similar to fundamental niches. The validity of this assumption is the key issue for interpreting resource partitioning. There are grave difficulties in describing patterns and inferring processes without considering alternative processes which could generate the same patterns (for example, Connor and Simberloff, 1979; Shipley and Keddy, 1987; Chapter 4). Alternative models and competing hypotheses can be produced by considering the possible differences between realized and fundamental niches. Figure 3.12 shows two possible responses to a removal experiment contrasting realized with fundamental niches. They assign fundamentally different roles to competition.

The interpretation of resource partitioning often has Panglossian overtones. The fictional Professor Pangloss asserted that whatever happens in this world, it is the best of all possible worlds. In the case of the Lisbon earthquake, which killed more than 30 000 people in 1755, he comments '... all this is for the best; for if there is a volcano in Lisbon, it could not be elsewhere; for it is impossible that things are not where they are; for all is well' (Voltaire, 1759).

3.5.2 Mechanism 1: differentiation of fundamental niches

One model of process assumes that resource partitioning results from different fundamental niches. The differentiation of fundamental niches is assumed to be the result of past competition selection for specialization. Each organism is

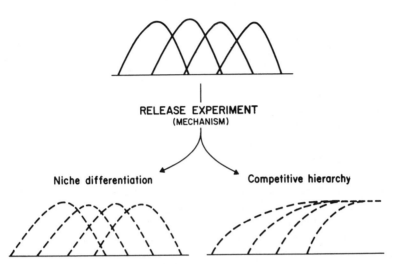

Figure 3.12 Two competiting models to account for resource partitioning. The top presents field observations (realized niches and physiological response curves) and the bottom presents the possibilities of differentiated fundamental niches (left) and competitive hierarchies operating on inclusive fundamental niches (right). Only experiments can distinguish between the two lower models.

specialized to harvest a particular region of the resource continuum for the obvious reason that 'a jack of all trades is a master of none' (MacArthur, 1972; Rosenzweig, 1979; Pianka, 1983). MacArthur observed that '... since competition often puts a premium on efficiency, this assumption implies a division of labor among specialists. It is the ultimate reason we have so many species'.

One of the most well-developed areas of enquiry using this model addresses the amount of permissable overlap in resource utilization by adjacent species. MacArthur (1972) showed that there were good theoretical reasons for expecting the maxima of adjacent resource utilization curves to be separated by $\sqrt{\sigma}$, where σ is the standard deviation of the curves (see also Schoener, 1974). Some of the arguments for such coevolution of competitiors are summarized by Roughgarden (1983). The observation that pairs of similar species tend to differ by ratios of from 1.2. to 1.4 (Hutchinson, 1959) actually predates such models by more than a decade, so the models do not so much predict ratios of limiting similarity as explain why they might be found. Simberloff (1983b) provides a critical review of the problems involved in exploring real data to test for unusually large differences in size ratios of coexisting species.

Such models assume that competition is relatively unimportant at present, serving only to produce minor differences between realized and fundamental niches (Fig. 3.12, left). That is, we cannot detect competition today because evolution has produced patterns of resource use which minimize interspecific competition. This view has been called the 'ghost of competition past' (Connell, 1980). Although it provides us with a convenient explanation for niche differentiation, and one that fits nicely with the theme of this book, its existence is rather more difficult to demonstrate. Connell (1980) proposed a rigorous series of experiments which need to be done to demonstrate that the ghost was there, but community ecology and competition research are still haunted by mere assertions of its presence. The first published example of such an experiment appears to be Turkington and Mehrhoff (1989).

A second common assumption is that the zones of overlap in resource use are a measure of competition, such that the greater the zone of overlap between two species is, the greater the intensity of interspecific competition (MacArthur, 1972; Schoener, 1974; May, 1974, 1981). This would be very convenient if it were true, and this view is probably very popular because 'competition' could be measured from descriptive data. It has been widely used to estimate competition coefficients in the community matrix, and has been defended recently by Schoener (1983). It is also wrong. Overlap in resource utilization curves tells us nothing about the intensity of interspecific competition, and it would be a serious error to construct a community matrix estimating competition coefficients from niche overlap. There is a simple *reductio ad absurdum* for this approach. Imagine two species sharing the resource continuum in the complete absence of competition: perhaps a predator is keeping population density very low, perhaps they are limited by a second resource. Since there is no competition their distributions merely reflect their fundamental niches, and overlap in the resource utilization curves, even if it is extensive, is occurring in the absence of competition. Now assume that two species are competing intensely. Since one of the tenets of the model is that competition reduces the width of resource utilization curves, the competition is so intense that there is complete competitive exclusion and no overlap whatsoever. Compare these two situations. In the situation with no competition there is extensive overlap, and in the situation with intense competition there is no overlap. Both of these situations are entirely consistent with the mechanisms normally postulated for the model, yet they demonstrate that niche overlap is smallest where competition is most intense. A similar point was made in the model describing habitat selection discussed above. Only controlled experiments can determine whether fundamental niches are differentiated in the manner assumed. The importance of such distinctions becomes clear when one of many possible competing hypotheses is considered.

Another important criticism is provided by Siefert and Siefert (1976). They note that if competition coefficients are estimated from descriptive data, they can take situations of overlap owing to mutualism or symbioisis and measure them as competition! This reinforces the point that studies of niche overlap may 'find' competition only because it was already assumed to be there. The tendency to ignore possible mutualisms in favour of competition is explored further in Chapter 8.

3.5.3 Mechanism 2: the competitive hierarchy model

The competitive hierarchy model proposes an alternative series of mechanisms to account for the observed pattern of resource partitioning and the differential distribution of species along environmental gradients. This model has been implicit in varying degrees in studies from a wide range of systems (for example, Connell, 1961, 1972; Miller, 1967; Sharitz and McCormick, 1973; Mueller-Dombois and Ellenberg, 1974; Colwell and Fuentes, 1975; Rabinowitz, 1978; S. D. Wilson and Keddy, 1986b) but has not been formalized to the degree that is desirable. There are three assumptions made by the model. First, it is assumed that the species in the community have inclusive niches; i.e. the gradient is a gradient of resource quantity, with all species having best performance (size, growth rate and reproductive output) at the same end of the gradient. Miller (1967) and Colwell and Fuentes (1975) provided many examples of this, and it may be the commonest situation for plants, which all share a requirement for a few basic resources: light, water and mineral nutrients. A second assumption is that the species vary in competitive ability in a predictable manner and that competitive ability is an inherent characteristic of a species, perhaps having something to do with rates of resource acquisition and capacity to interfere with neighbours. Lastly, it assumes that competitive abilities are negatively correlated with fundamental niche width, perhaps because of an inherent trade-off between ability for interference competition and ability to tolerate low resource levels. This is illustrated in Fig. 3.13(top), where six species are ranked in competitive ability in order from A (dominant) to F (subordinant). The outcome is species differentially distributed along a gradient (resource partitioning), as shown in Fig. 3.13(middle). However, the mechanism is a dominance hierarchy with the competitive dominant occupying the preferred end of the gradient, and the subordinants displaced down the gradient a distance directly determined by their position in the competitive hierarchy (see also Fig. 3.12).

Unlike the resource partitioning model, the competitive hierarchy model is predictive. Given a knowledge of fundamental niches or competitive abilities in the preferred region, one can predict the order in which species will be distributed along the gradient. This assumes, of course, that competitive ability is an inherent trait of a species rather than a trait which is strongly dependent upon the environment. If this is the case, then measuring competitive ability should allow us to predict the distribution of organisms in such communities.

There is another interesting parallel here. At the beginning of this chapter we compared the Lotka–Volterra model with the resource competition model,

FUNDAMENTAL NICHES

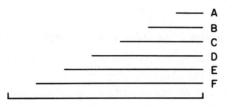

REALIZED NICHES: STRICT ASSUMPTIONS

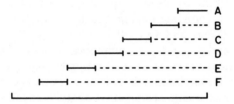

REALIZED NICHES: RELAXED ASSUMPTIONS

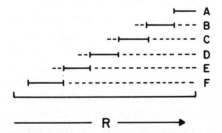

R ⟶

Figure 3.13 Aspects of the competitive hierarchy model. (top) The fundmental niches of six species with competitive abilities arranged in a hierarchy from the dominant A to the subordinant F. (middle) The observed field distributions of the species in the top panel assuming that in each interaction the dominant excludes the subordinant right to the very limits of its physiological tolerance limits. The broken lines illustrate the species distributions which would be detected by a competitive release experiment. (bottom) Relaxing an assumption. Near its tolerance limits each dominant is competitively excluded by its adjacent subordinant. The distribution patterns remain the same except that each species is displaced slightly up the gradient. The broken lines showing the results of a release experiment emphasize that competition has very different effects on distributional limits of species depending upon whether the limit is adjacent to a dominant or a subordinant.

and found that although they both yielded similar predictions, the latter model was preferable because it specified the mechanism of interaction. With respect to resource partitioning, we have a similar set of circumstances: two models which generate the same outcome, but one placing more emphasis on the mechanism.

3.5.4   A variant on the competitive hierarchy

In its simplest form it is assumed that the transition from dominant to subordinate occurs at the exact point where the dominant reaches its physiological tolerance limits (Fig. 3.13, middle). This would be true in the cases of absolute asymmetry; that is, where the dominant affects the subordinate, but the subordinate has no effect upon the dominant. Suppose that this assumption is relaxed slightly. The consequence is that the transition from the dominant to the subordinate occurs only near the lower distributional limits of the dominant, since there is presumably some point at which the dominant is so weakened by environmental effects that it can be excluded by the subordinate (Fig. 3.13, bottom). Another way of expressing this is to say that we have relaxed the assumption that for each dominant–subordinate interaction the realized and fundamental niches of the dominant must be identical. Depending upon how far one relaxed this assumption, one could produce a series of cases intermediate between resource partitioning and strict competitive hierarchies. The order of species distributions would remain identical, but the distributions of species would shift slightly up the gradient towards the preferred end.

In this case competition is again affecting both ends of a species' distribution, but it plays a major role at one end and a minor role at the other. Field experiments may have to take this possibility into account. The essential issue in distinguishing between these two possibilities becomes the behaviour of the dominant–subordinate interaction near the distributional limits of the dominant.

3.5.5   Centrifugal organization of communities

This competitive hierarchy model and the foregoing model of behaviour and habitat use (Pimm and Rosenzweig, 1981; Rosenzweig, 1981) are closely related, and suggest a general model of community organization. Rosenzweig and Abramsky (1986) have recently extended their model of habitat use by proposing a type of community structure termed **centrifugal organization**. In such situations, a group of *n* species have shared preference for a central habitat type, but each has a another peripheral habitat in which it is the best competitor; the number of different peripheral habitats then determines the number of species which can coexist. This is a variant on the more usual form of inclusive niche structure (Miller, 1967; Colwell and Fuentes, 1975) where species have overlapping fundamental niches along only one axis. Keddy

(1989a) has extended the centrifugal model to more complex communities by postulating that not just single habitats, but entire environmental gradients (or niche axes) may radiate outward from the central preferred habitat, accommodating many more species than in the Rosenzweig and Abramsky version (Fig. 3.14). Near the centre species may have entirely inclusive fundamental niches, but nearer the periphery, species' fundamental niches may include only a few adjacent neighbouring species in the direction of the central habitat. In the latter case competition would be simply 'one sided' and removal experiments would then be predicted to show that in the absence of neighbours, species could grow nearer the central habitat but not nearer the periphery. There is evidence

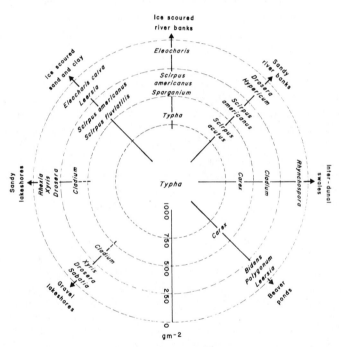

Figure 3.14 Competitive hierarchies along different biomass gradients produce centrifugal organization of wetland plant communities. Species have inclusive niches along each axis with shared preference for the fertile, undisturbed central region. This region is dominated by a few canopy-forming, clonal perennials, primarily *Typha* sp. (after Keddy, 1989a.)

that wetland plant communities are organized in this fashion (Keddy, 1989a). In wetland plants the central habitat has low disturbance and high fertility, and is dominated by large leafy species capable of forming dense canopies. Different constraints, such as kinds and combinations of infertility and disturbance, create radiating axes along which different groups of species and vegetation types are arrayed. Rare species occur only in peripheral habitats with low biomass (Moore *et al.*, 1989). Whether such a model is valid beyond desert rodents (Rosenzweig and Abramsky, 1986) and wetland plants (Keddy, 1989a) needs to be explored.

3.5.6   Comparison of resource partitioning models

That different populations use different resources and habitats is basic natural history. What is less clear are the mechanisms which underlie these patterns. The resource partitioning model places emphasis upon the mechanisms of differentiated fundamental niches, whereas the competitive hierarchy model places emphasis upon proximal competitive interactions. The resource partitioning model assumes that evolution, to avoid interspecific competition, has produced different fundamental niches, whereas the competitive hierarchy model assumes that there are inherent trade-offs between fundamental niche width and competitive ability. The evidence for evaluating the biological reality of these models comes from two very different sources. Those who use descriptive data tend to assume that resource partitioning comes from different fundamental niches, whereas those who conduct field experiments often find evidence of competitive hierarchies. It would be tempting to assume that the former group does not find inclusive niches because they never look for them. For example, experimental studies of nectivorous birds illustrate competitive hierarchies; (Pimm, 1978) as do descriptive studies (Pimm and Pimm, 1982); Willer (1967) suggests other examples.

This illustrates an important point about the value of models, and the potential for both use and abuse. If the model is treated as a demonstrated truth, then there is always the temptation to collect yet another set of data demonstrating the existence of this truth. There is no doubt that communities have populations with different realized niches (Schoener, 1974; Giller, 1984) as the model illustrates. However, this is no reason to assume that the mechanisms presumed to underlie the model, particularly the similarity of realized and fundamental niches, are correct. Competing hypotheses need to be advanced and tested. The same model can therefore act either as a hindrance or as a stimulator to the advancement of science, depending upon how it used as a research tool.

A problem in evaluating models occurs when different models with different assumptions make identical predictions (B. Shipley, pers. commun.). For example, the observation that the number of species in a community reaches a maximum at some intermediate level of disturbance is well recognized, but there are different mechanisms proposed to account for it (for example Grime, 1973, 1979; Connell, 1978; Huston, 1979; Tilman, 1982). The foregoing models therefore cannot be evaluated by comparing their predictions, because their predictions are the same. In such cases the best research strategy appears to be comparing and contrasting the assumptions of the models and designing experiments to compare the validity of these assumptions.

As an alternative, what we may need is not critical tests so much as a resolution. Some communities may have populations with different fundamental niches, whereas others may have inclusive fundamental niches. For example, Yodzis (1978, 1986) postulated that competition for 'resources' is fundamentally different from competition for space, the former having symmetric and the latter asymmetric interactions. Another possibility is that communities with strong resource gradients (such as salt- and freshwater shorelines) evolve very different structuring from communities occupying relatively homogeneous habitats (old fields or prairie parkland); of course, homogeneity is in the eye of the beholder, so it may similarly be that sessile organisms generally experience strong localized gradients, whereas birds and mammals tend to experience relatively more-homogeneous conditions. Lastly, it may be important to consider the kind of gradient itself. If there are gradients of resource quantity (food abundance, moisture and soil nutrients), they may be fundamentally different from gradients of resource quality (kind of food and ratios of nutrients). The resulution is therefore likely to require explicit consideration of the kinds of resources and their distributions in nature. Meanwhile, ecologists cannot automatically assume that all communities fall conveniently into one model.

3.6 CONCLUSION

The foregoing models illustrate exploratory, summary and, to a lesser extent, predictive models used in the study of competition. It is not yet clear whether any of them will provide a foundation for a solid body of competition theory. One route forward may lie in carefully designed mechanistic models based on realistic assumptions. Experimentalists could then interact with modellers by testing the assumptions used to construct the models, and testing whether these assumptions generate the expected predictions (for example, Austin, 1986). It will probably be less useful to succumb to the temptation to refine existing models endlessly in the hope that increased complexity of mathematics will generate a closer approximation to reality. Going by this route there is a genuine risk that models will be twisted to make them explain everything, rather than them serving as clear signposts and reference points.

Mechanistic models may be unable to attain the accuracy of prediction provided by simple correlation models; the quantitative description of general ecological patterns may therefore be the goal for which we ought to strive (for example, R. H. Peters, 1980a; Rigler, 1982). The advantage of this approach is that it provides an easy measure of the validity of a model – r^2. The higher the percentage of variation in nature accounted for by the model (as measured by r^2), the more useful the model. Models such as this were not examined in this chapter because they do not deal with mechanisms such as competition, only with the resulting patterns such as the relationship between algal biomass and dissolved phosphorus in lakes.

However, such correlational approaches to the study of competition can be expanded in a mechanistic direction. One of the best examples comes from the studies of 'self-thinning' in plant monocultures referred to in Chapter 2 (Fig. 2.4). There is a general and well-established relationship between the mean mass of individuals and the density at which they are grown (Harper, 1977; Westoby 1984) with a slope of $-3/2$. This empirical relationship (which is often known as the '$-3/2$ law') has broad generality, in that it can be applied to many kinds of plants grown both in the field and under laboratory conditions (Harper, 1977; Gorham, 1979) and can be clearly related to principles of geometry (Whittington, 1984). Thus, simple models already exist to predict plant performance from density, and there is little doubt that the principal mechanism is intraspecific competition for limiting resources. Such empirical approaches could be expanded and applied to other areas of ecology. For example, Austin and Austin (1980) and Austin (1982) have sought relationships among plant traits such as physiological performance, morphology and position along experimental nutrient gradients. S. D. Wilson and Keddy (1986a) suggest the way in which we could produce regression models predicting competition intensity in herbaceous vegetation. More recently, Gaudet and Keddy (1988) have shown that plant competitive ability can be predicted from plant traits. There is much potential for developing such simple predictive models but, whatever their advantages, some scientists will probably continue to prefer models which are mechanistic (for example, Tilman, 1987a).

At present the greatest impact of models upon the study of competition has not been the accuracy of their predictions or the realism of their assumptions. Rather, they have provided the context or setting which defines the sort of questions that are thought to be interesting. The wealth of studies in resource partitioning illustrates the way in which theory guides observation. The next chapter examines more closely how concepts and models can be evaluated using data from real ecological systems.

3.7 QUESTIONS FOR DISCUSSION

1. What are the benefits of constructing ecological models?

2. Are there any ways to overcome the inherent trade-off between precision and generality? What do we mean when we say that combining the two depends upon the skill of the modeller?

3. Why has resource partitioning had such an impact upon the ecological literature?

4. What should be the objectives for the next generation of ecological models? Can we specify criteria which these models should satisfy?

5. Compare and contrast the benefits of mechanistic models as opposed to simple predictive models.

6. Is Panglossianism a philosophical attitude or a falsifiable hypothesis for the nature of the world?

7. What kinds of models would be of most use in maintaining the biological diversity of our biosphere? Are they likely to have anything to do with competition?

4 Choosing the tools

Give us the tools and we will finish the job.

Winston Churchill, 1941

The mere taking of an instrument into the field and
recording of observations, or the collection and
analysis of... samples, is no guarantee of scientific
results.

A. G. Tansley (1914)

...there are never any guarantees of success and one
always remains at the mercy of triviality or poor
judgement. No methodological principle can eliminate
the risk... of persisting in a blind alley of
inquiry.

I. Prigogine and I. Stengers (1984)

In order to answer new questions, or to develop new techniques, it is necessary
to consider the tools which ecologists have for exploring nature. Let us begin
this chapter with a few general observations on the use of tools. Any
craftsperson needs to begin his or her craft with a clear understanding of the
uses and abuses of each tool. It is, however, possible to continue using an
unnecessary tool simply because it is easier to do so than to learn how to
use a new (and different) one. New tools can be invented once the old ones
have been mastered, but new tools are usually invented to solve new problems. Finally, tools do not solve problems or build things by themselves:
they are guided by a blueprint. No amount of skill with tools can compen-
sate for a sloppy or incorrect blueprint.

The blueprint for ecologists is the research strategy, and the tools are
conceptual approaches. Much of the challenge in current ecological research
lies in deciding which research strategies or paths are the most efficient. This in
turn raises questions about what forms of scientific evidence are admissible,
and how these forms are to be properly interpreted. Several conceptual
approaches to the study of competition currently coexist (albeit uneasily), and
there is no general agreement yet upon their value (for example, Saarinen,

1982; Quinn and Dunham, 1983; Simberloff, 1983a). These conceptual approaches can be applied to any area of ecological research. If new conceptual approaches are to be developed, then there must be familiarity with and understanding of existing conceptual approaches. This is the foundation for innovation.

Two questions will serve as the theme of this chapter. (1) How do we investigate nature to discover and measure competition? (2) What are the relative merits of these different approaches? We could rephrase this as: 'What tools currently exist, and how do we look for new ones?'. The chapter concludes by stressing that tools are only that – they must be skilfully used. Skilful use includes beginning with a clear question, choosing an appropriate model system, and justifying the validity of proposed tests *before* data are collected and analysed.

4.1 DESCRIPTIVE, COMPARATIVE AND EXPERIMENTAL STUDIES

There are three basic approaches which have been used to explore the importance and role of competition in nature. The first is descriptive, where data sets are collected from a community and are then statistically manipulated to look for patterns which the researcher believes can be attributed to competition. This idea was well developed in plant community ecology, and it has been popularized by animal ecologists within the past decade.

The two remaining approaches depend upon comparing different communities to answer one or more specific questions. In comparative studies the independent variable is not under the researcher's control. An example would be the comparison of phytoplankton communities in lakes where acid rain has created different pH levels. In experimental studies the researcher deliberately manipulates one or more independent variables in some communities, while keeping others as unmanipulated controls. An example would be a removal experiment, where a postulated competitive dominant is removed from treatment plots but left alone in control plots.

Diamond (1983, 1986) provides an overview of these kinds of comparative and experimental approaches, although he uses a different classification. Diamond suggests that ecologists must draw upon disciplines such as geology and astronomy, rather than on chemistry or molecular biology, when considering the methods for investigating ecological systems. He recognizes three kinds of experiments, which are regions along a continuum of possibilities: laboratory experiments, or perturbations produced by an experimenter in a laboratory; field experiments, or perturbations produced by an experimenter in the field; and natural experiments, where the experimenter exploits naturally occurring perturbations. Each kind of experiment is appropriate for answering different questions at different spatial and temporal scales. Within the category of 'natural experiment' Diamond recognizes a further subdivision. Natural trajectory experiments are comparisons of the same community before, during and after a perturbation produced by nature or by humans other than ecologists. Examples would include fires or insect eruptions in a forest, floods, storms and introductions of exotic species. Natural snapshot experiments are comparisons of communities assumed to be different principally with respect to one independent variable. Examples would include comparisons of grassland areas with and without herbivores, or insect communities in polluted versus unpolluted water courses. Such studies always depend upon a convincing demonstration that only one important independent variable differs between the communities being compared.

Diamond then evaluates the strengths and weaknesses of each of these four types of experiments according to eight criteria (Table 4.1). There are clear trade-offs. In general, the larger the spatial or temporal scale being inves-

Table 4.1 Comparison of strengths and weaknesses of different types of evidence on the importance of competition in structuring ecological communities (modified from Diamond, 1986, Table 1.1)

| | Source of evidence | | | |
| | Experimental study | | Comparative study[a] | |
Axes	LE[b]	FE[c]	NT[d]	NS[e]
Control of independent variables	H	M/L	–	–
Matching of sites	H	M	M/L	L
Maximum temporal scale	L	L	H	H
Maximum spatial scale	L	L	H	H
Range of possible manipulations	L	M/L	M/H	H
Realism	L	H	H	H
Generality	–	L	H	H

[a] Diamond calls these 'natural experiments'; [b] laboratory experiments; [c] field experiments; [d] natural trajectory studies; [e] natural snapshot studies.

tigated is, the less control there is over independent variables and site matching. A researcher must specify in advance which scale is being used or, better still, should use a research strategy employing several of these approaches simultaneously.

Diamond's classification can be criticized for stretching the word 'experiment' beyond its normal scientific meaning. In normal usage, an experiment has control plots and treatment plots, with the experimenter in full control of the independent variable that is initially changed. Since 'experimental ecology' is now popular, the desire to put descriptive field studies into the experimental category is understandable. However, it runs the risk of muddying the term 'experimental'. At the same time natural experiments, if they have clearly stated questions, are clearly different from simple descriptive studies where patterns are sought and interpreted after the data are collected.

The word experiment should probably be used more narrowly. In this case three classes of study can be identified: descriptive, comparative and experimental (Fig. 4.1). The term comparative would apply to both kinds of natural experiments, and emphasizes the principal trait of this kind of study: comparison of two or more systems separated in space or time, where the systems are selected by the scientist rather than created.

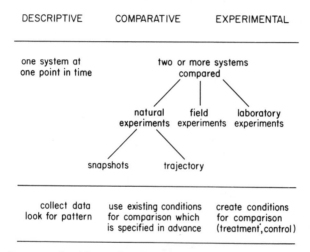

Figure 4.1 Sources of evidence for the existence of competition (a revision of Diamond, 1983, 1986). Table 4.1 lists some strengths and weaknesses of the experimental and comparative approaches.

4.2 DESCRIPTIVE STUDIES

The description of pattern and inference about competition is best illustrated using plant communities, since generations of plant ecologists have been occupied with tallying the contents of quadrats in the summer, and then trying to draw inferences about these observations in the winter. An impressive array of statistical techniques has developed for looking for such patterns (for example, Kershaw, 1973; Mueller-Dombois and Ellenberg, 1974; Orloci, 1978; Pielou, 1977, 1984; Gauch, 1982). Fowler's (1986) review of competition in arid areas and the exploration of zonation patterns and competition on shorelines by Shipley and Keddy (1987), illustrate the close relationship which can exist between pattern analysis and questions about competition.

Since we are interested not in mathematical sophistication but in inferring competition from pattern, we can take the simple example of association analysis using the 2×2 contingency table. In this approach data are collected from *n* sample units (quadrats or other sampling devices) and the association between any pair or species is calculated using the chi-square test (for example, Siegel, 1956; Kershaw, 1973; Pielou, 1977). The null hypothesis is that within the *n* sample units the species are independently distributed. The alternative hypothesis is that the two species are either positively or negatively associated. If this procedure is carried out for each pair of species in the data set, a constellation diagram can be drawn to illustrate either positive or negative associations within the community of interest. This simple technique provides a visual summary of the patterns of species within a particular community (Fig. 4.2). In general it is positive association which is mapped in such figures, but negative association could be mapped in the same way. It is

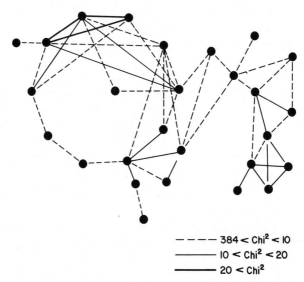

$$------\ 384 < Chi^2 < 10$$
$$————\ 10 < Chi^2 < 20$$
$$————\ 20 < Chi^2$$

Figure 4.2 A typical constellation diagram showing positive association among 27 shrub species in Californian coastal sage scrub (after Kirkpatrick and Hutchinson, 1977). Of course, negative associations rather than positive ones would be needed to be plotted if we wished to find patterns consistent with competition. In either case, however, there are so many plausible mechanisms consistent with such patterns that these descriptions do not allow us to infer mechanisms, but only generate hypotheses regarding them.

patterns of negative association which are of most interest in studies of competition, but interpreting the causes of such patterns is much more difficult. There are at least three possibilities, depending upon how broadly they are defined. Four are listed below.

1. The species are restricted to different microhabitats, and thus do not interact. This may be the result of different adult physiologies, or much less obvious differences at the regeneration stage (Harper, 1977; Grubb, 1977);

2. The species are, in fact, positively associated, but the sample unit was so small that only a few individuals fit in it, thereby obscuring the pattern which occurred at a slightly larger scale;
3. Agents such as predators independently control each species and restrict each to a different set of conditions. Such circumstances could fall within case 1, depending on how broadly it is conceived;
4. The species compete, and competition leads to habitat separation.

It is not possible to distinguish among these causes with descriptive data alone. Hypothesis 1 could be falsified in theory if one could demonstrate that the species in fact had identical physiological tolerance limits and identical regeneration requirements. In practice this is not feasible, because even if the species were shown to be absolutely identical along n axes there is always the possibility that they differ along the $(n + 1)$th. Characteristics such as requirements for mycorrhizal fungae or obscure insect seed vectors are easy to overlook. Hypothesis 2 could be eliminated in theory by using sample units of many different sizes and demonstrating that the negative association is not merely a result of inappropriate choice of quadrat size. Hypothesis 3 has the same practical difficulties as hypothesis 1, and thus cannot be falsified.

If the alternatives cannot be falsified, then we cannot demonstrate that competition is the only plausible mechanism. This message has been widely accepted in plant ecology at least, so that negative associations are (usually) accepted no longer as evidence for competition. The value of such tools is that they suggest hypotheses which can be tested experimentally.

A variant on this approach is to choose natural environmental gradients and examine the distributional limits of species along these gradients in order to infer the existence of competition. The basic assumption is that systems that are structured by competition have different kinds of patterns from those which do not. For example, some studies have described the patterns of distribution of trees (Whittaker, 1956, 1967) and birds (Terborgh, 1971) along altitudinal gradients. The problem with interpreting such data is that it is not clear what sort of patterns competition would produce, or whether the observed patterns were statistically significant. Pielou (1977, 1979), Underwood (1978) and Shipley and Keddy (1987) have proposed statistical tests for comparing distributions of species along gradients; three alternatives can be recognized.

1. Species distributional limits are regularly spaced along the gradient like tiles on a roof;
2. Species distributional limits are randomly arranged along the gradient;
3. Species distributional limits are clustered along the gradient, producing apparent communities.

These techniques have recently been applied to freshwater shoreline communities (Keddy, 1983), and yield a number of observations on pattern: (1) species distributions are clustered along the gradient; (2) upper and lower distributional limits differ in the intensity of clustering; and (3) other factors such as disturbance from waves change the degree of departure from the null model. Figure 4.3 illustrates all three of these observations using data from the shore of a small lake.

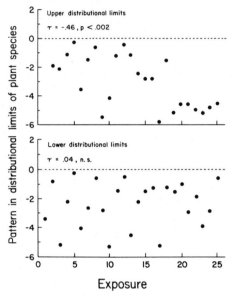

Figure 4.3 Another test for the non-random distribution of species, in this case one exploring the distributions of species along natural environmental gradients (for example, Pielou, 1977). This example, the distributions of plants along a water depth gradient on a lakeshore (after Keddy, 1983), starts with the null model that species distributions are randomly distributed along this water depth gradient (hatched line), and tests whether either the upper or lower limits of species distributions deviate from it. Both do ($P < 0.001$). For upper, but not lower, limits the degree of departure from the null model increases along a second gradient, exposure to waves ($P < 0.002$). Exposure incorporates increasing disturbance and decreasing fertility from left to right. Do such patterns tell us anything about competition?

Assume now that we have a system with a null model, and that we have demonstrated departures from the null model as in Fig. 4.3. What, if anything, does this tell us about the nature of competition? Again, the answer has to be virtually nothing. We could devise scenarios where interspecific competition produced departures from the null model towards either regular or clustered patterns. So the existence of non-random distributions *per se* does not demonstrate that interspecific competition is producing the observed discontinuities. (On the other hand, it does tell us that there is a real pattern rather than an imagined one, and does provide a tool for testing whether such patterns are of general occurrence before we design experiments to investigate them further.) Consider the hypotheses to account for clustered distributional limits.

1. The species may have similar distributional limits because of similar physiological tolerance limits. All may have a particular mechanism for tolerating flooding, in which case they all stop at the same position along the gradient;
2. Clusters of distributional limits may be attributed to the way in which the observer divided up the gradient. In Fig. 4.3 the shoreline was divided into 5 cm elevation ranges using an automatic level; quadrat boundaries set by, say, soil oxygen concentration, may have yielded entirely different patterns;
3. Grazers such as moose or muskrats may stop at a certain water level, thereby creating a discontinuity;
4. One or more competitive dominants may set the distributional limits of an entire group of species.

Only hypothesis 4 is consistent with competition. As with the example of association analysis, hypotheses 1–3 would be difficult to reject conclusively, since one can always postulate the existence of new, unexplored environmental factors. In the case of Fig. 4.3, Keddy (1983) showed that the upper distributional limits of species coincide with the presence of woody plants. This was the strongest evidence of competition that could be drawn from such descriptive data (the resulting field experiment where shrubs were manipulated is described in Keddy (1989b)).

Although the techniques of association analysis and its modern multivariate counterparts differ from those of direct gradient analysis, each has been used to search for patterns attributable to competition. In both cases the outcome is that no amount of description of pattern can demonstrate the existence of competition, since alternative equally plausible hypotheses can always be erected to account for the same observations.

4.2.1 Assembly of species communities and null models

Plant ecologists have not been the only group to confront the problem of inferring competition from pattern. In his paper entitled 'Assembly of species communities', Diamond (1975) describes the distributions of bird communities in New Guinea and its satellite islands. In this case, instead of quadrats, islands are the sample units. For each island there was a list of species observed, and Diamond used these data to define rules for the assembly of bird communities. The assembly rules described the patterns in this data set, and Diamond then inferred that factors such as differing dispersal abilities and competition among species of birds were the primary cause of these patterns. For example, he divided the bird fauna into six groups based on their tendency to occur on islands ranging from species-rich to species-poor. Supertramps were the group which tended to occur only on islands with low bird species richness. Such species could be considered fugitive species with a strategy of 'breed, disperse, tolerate anything, specialize in nothing' (Diamond, 1975). The evidence for competition was primarily based upon pattern. Certain pairs of species did not co-occur on islands where they might otherwise be expected (Fig. 4.4) or, where they did co-occur, their within-island distributions were different from those on islands where they occurred without the other member of the pair. These observations were supplemented by observations on habitat and food requirements.

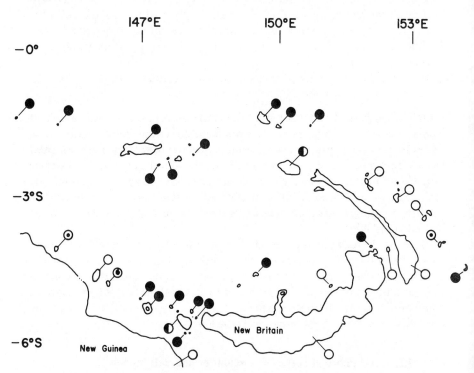

Figure 4.4 The distributions of two species of *Ptilinopus* fruit pigeons in New Guinea (split circles indicate co-occurrences and dots co-absences; after Diamond, 1975) Pictures of such 'checkerboard distribution patterns' lack the statistical rigour of association analysis (Fig. 4.2) or gradient analysis (Fig. 4.3), but they are often used as evidence of non-random distributions. The assumed negative association of such species is frequently attributed to competition.

Diamond's assumption of competition was harshly challenged by Connor and Simberloff (1979). A major criticism was the lack of a 'null model' against which species distributions could be compared. What would we expect the distributions to be in the absence of competition, and how would we test whether the observed patterns deviated significantly from that expectation? The search for an appropriate method of constructing null models has continued for nearly a decade. The major unresolved question is how to construct such a model. Are elements of biological reality necessary? If so, which ones? Do they bias the test in a direction favouring or weakening the detection of competition? (See, for example, Connor and Simberloff, 1979; Grant and Abbott, 1980; Diamond and Gilpin, 1982; Gilpin and Diamond, 1982; Wright and Biehl, 1982; Simberloff, 1983a, 1984). What seems to have been overlooked is that no amount of statistical sophistication will provide unambiguous evidence of mechanisms from descriptions of pattern. It is interesting that the above studies do not refer to the extensive literature on pattern analysis in plant communities, particularly the later work in which specific null models were proposed (for example, Pielou, 1977, 1979; Underwood, 1978; Dale, 1984). The presence of clear null models for species distributions along environmental gradients provided a tool to test whether the observed patterns were statistically significant, but left plant ecologists as uncertain as ever as to their cause (Shipley and Keddy, 1987). There is every reason to believe that the null models and island patterns issue, although having generated controversy (and not a small amount of simple rudeness), will lead to exactly the same unsatisfactory conclusions that were reached above with much simpler null models. Even if a consensus is reached on the unambiguous construction of ecologically meaningful null models, we will still have to resort to other sources of evidence to know that causes these patterns.

4.3 COMPARATIVE STUDIES

Comparative studies follow directly from descriptive studies, in that observational data are used to describe patterns, and then the resulting patterns are compared in order to infer differences in process. The comparison of patterns from different spatial or temporal locations defines this conceptual approach; examples include comparing between two islands or habitats, or exploring variation in pattern along environmental gradients. Given the descriptive examples cited above, there is an obvious area where these approaches intergrade.

The value of comparative studies lies in the spatial and temporal scales which they can consider (Table 4.1). MacArthur (1972) argued that because competition may occur only infrequently, direct measurement may be difficult. He proposed that inferences from biogeography will therefore provide most evidence for competition. However, such sources of evidence should be used with caution. First, because there is not an experiment, one is always forced to compare patterns and then invoke mechanisms (for example, Diamond, 1975, 1986). When many hypotheses make identical (or at least similar) predictions about pattern, this process simply does not work. Given the complexity of nature, it is almost always possible to come up with alternative hypotheses, as the preceding section on descriptive data illustrated.

Two examples will illustrate comparative approaches. The first example comes from lichen communities, where, because of slow growth, experiments are very difficult. Lawrey (1981) tested for competition in lichen communities by comparing two islands. Because of air pollution, one island had many fewer species. This is therefore a snapshot study (Fig. 4.1). Lawrey predicted that competition would be less intense on the polluted island, and used four kinds of evidence to test this prediction (Table 4.2). The first and second sources of evidence come from spatial distribution, which has already been shown to be

Table 4.2   Sources of evidence of competition used in comparative studies. Example 1, illustrating tests of pattern applied to lichen distributions on islands ($n = 2$ species, *Xanthoparmelia conspersa* and *Pseudoparmelia balitmorensis*; data from Lawrey, 1981)

Pattern	Results	Interpreted as evidence for competition?
Distance between thalli	Thalli closer on unpolluted island	Yes
Identity of nearest neighbour	More-frequent interspecific associations on unpolluted island	Yes
Niche breadth	Greater on polluted island (one species)	Yes
Niche overlap	Greater on polluted island (one species)	Yes

inappropriate for demonstrating the existence of competition. Increased realized niche breadth is certainly more convincing, because fewer alternative hypotheses for this change can be postulated. However, the very fact that one island is polluted suggests many other hypotheses that could account for a change in a species distribution. The difficulty in interpreting changes in niche overlap is dealt with in Chapter 3. There is no doubt that patterns of lichen distribution are different on a polluted and non-polluted island, but competition is only one of many hypotheses which could account for it.

It is also difficult to do experimental studies of birds on islands. Moreover, Moulton and Pimm (1986) argue that experimental studies of small sets of species are less useful than studies using many species simultaneously. They therefore examined patterns of extinction of introduced bird species on the Hawaiian islands. In contrast with Lawrey, their work illustrates a trajectory study. Instead of exploring two sites differing in space, they have data collected from one site over time (although some between-island comparisons are also made), and instead of comparing patterns of distribution they compare species morphologies. Moulton and Pimm use a battery of tests to look for patterns which might be attributable to competition (Table 4.3). The basic argument

Table 4.3   Sources of evidence of competition used in comparative studies. Example 2, illustrating the array of tests applied to birds on the Hawaiian islands by Moulton and Pimm (1986)

Pattern	Results	Interpreted as evidence for competition?
Extinction rate versus species number	Extinction rate increases with the number of species on an island	Yes
Morphological similarity		
Coexistence of congenors	Bill length more similar in 9 pairs which have gone extinct than 6 which coexist	Yes
Morphological patterns in entire community	Morphological differences overdispersed relative to random communities	No
Probability of extinction	Probability of extinction higher in congeneric pairs which are more similar $(N = ?)$	Yes
Species to genus ratio	No evidence that species in same genus are more prone to extinction	No
Species abundance patterns	Species abundances lower on island with more species	Yes

underlying these tests is that similar species are more likely to compete, and therefore less likely to coexist. Although the tests do not all agree, Moulton and Pimm conclude that similar species are less likely to coexist. Assuming that one considers the patterns to be real, it is next necessary to ask whether competition produced them. Could alternative hypotheses be erected? None is offered, and it may be more difficult to generate reasonable alternatives than with the preceding study. However, it seems likely that any general pattern in communities could be accounted for by a series of competing hypotheses. When competing hypotheses are not listed and falsified, one can consider the pattern demonstrated but the mechanism is a matter of conjecture.

The difficulty in inferring mechanism from pattern is the principal difficulty with comparative studies and, in this sense, comparative studies are no different from descriptive studies. A second problem with the comparative approach is that, as in the descriptive approach, it may be tempting to collect comparative data before one even has a question. A field experiment requires one to specify a hypothesis and its test in advance; a comparative study can be a justification for the collection of an elaborate data set without any clear questions specified in advance. Because ecologists usually love working with real ecosystems, it is always tempting to pick up quadrats, binoculars, sample bottles or nets and head to the field without any questions at all.

4.4 EXPERIMENTAL STUDIES

In order to design an experiment, one must start with a question. The sorts of questions which laboratory and field experiments answer are fundamentally different. However, both begin with the hypothesis that competition may be structuring a community in a certain manner. A manipulation is then performed to test this hypothesis. Experiments have a number of well-defined steps (Hicks, 1964). These include: (1) choosing a question; (2) selecting dependent and independent variables; (3) designing control plots as reference points for the treatments; and (4) planning randomization to ensure that unexpected and uncontrolled forces do not appear as treatment effects.

One of the first steps in experimental design is to specify the independent and dependent variables which are to be used. In studies of competition the dependent variables of interest are usually traits which measure the performance ('fitness') of individuals or populations. The specific traits which are measured will depend upon the kind of organism being studied, but could range from simply the size of individuals to their distribution along an environmental gradient. Choosing the best dependent variable requires a real understanding of and familiarity with a system, and often using several simultaneously is best. For example, in his studies of tadpole competition, Wilbur (1972) used three dependent variables: the length of time to metamorphosis, survival rates and weight at metamorphosis. Each of these has obvious importance for amphibians completing their larval stage in temporary ponds.

The independent variable is the abundance of neighbours. To test for the presence of competition in an experiment, the researcher predicts a negative relationship between the abundance of neighbours and the performance measures.

4.4.1 Laboratory experiments

The objective of a laboratory experiment is quite different from that of comparative or field experimental work. Laboratory experiments tell us what could potentially happen in nature, and what might occur under specific sets of conditions which we cannot produce in the field. They have the advantage that the experimenter can manipulate a wide range of biotic and abiotic variables, but their principal drawbacks are extreme unrealism and extremely limited scope (Diamond, 1986). However convincingly one demonstrates competition in laboratory experiments, one cannot extrapolate from them to the actual existence of competition in nature.

Laboratory experiments are most important for testing whether certain postulated relationships could potentially occur (Fig. 4.5). Laboratory experiments are therefore discussed primarily in Chapter 6. The question being pursued at present is the actual role of competition in nature – that is, region 'a' of Fig. 4.5, and the two central columns of Fig. 4.1.

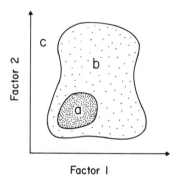

Factor I

Figure 4.5 Regions of enquiry for competition theory (and ecological theory in general). For any pair of factors (e.g. population sizes and environmental conditions) three regions can be recognized. Region a is that which actually occurs in nature. Region b does not occur in nature, but can be created in the laboratory. Region c cannot be created experimentally, but may still be of theoretical interest. Region a can be explored with field experiments. Region b, and perhaps parts of region a, can be explored using laboratory experiments. Regions a, b and c constitute the conceivable world, where c can only be studied using models.

4.4.2 Field experiments

In a field experiment, there is again the hypothesis of a negative relationship between performance and abundance of neighbours. Unlike laboratory experiments, however, there is a reference point: the current performance of the population(s) of interest (Fig. 4.6, solid dot). Two perturbations are then possible. If the abundance of neighbours is increased (an additive experiment, open triangles), then measures of performance are predicted to decline. If the abundance of neighbours is reduced (a removal experiment, open circles), then measures of performance are predicted to increase. In the absence of any measurable effect, the null hypothesis – no competition – must be accepted.

There are several additional considerations in such experiments. First, the relationship between performance and abundance will probably vary among habitats; Keddy (1981) has explored some of the possible patterns that such relationships may exhibit. The reference point in the experiment (solid dot) simply specifies one position along one curve in a family of curves. The objective of the experiment is to change the position along the curve without deflecting the habitat from one curve to another. Secondly, there are different

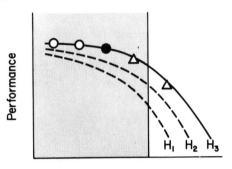

Abundance of neighbours

Figure 4.6 One can explore for competition experimentally by testing for a negative relationship between performance and abundance. Different habitats (H_1–H_3) may have different performance–abundance curves. The solid dot shows the present state of the system in habitat H_3; open circles are removal experiments and open triangles are additive experiments. If these manipulations produce no measurable change in performance, then there is no evidence for competition – i.e. the performance–abundance curve is assumed to have a slope of zero in the region of the manipulations. Additive experiments may test for competition either within the 'natural' range of abundance (stippled region) or in artifically high abundances. In the latter cases they approach laboratory experiments by examining hypothetical circumstances.

ways of manipulating abundance. Bender *et al.* (1984) draw a distinction between **press** experiments, where the density of neighbours is changed and held at a new level, and **pulse** experiments, where the abundance of neighbours is perturbed but then allowed to recover. They discuss some of the possible difficulties in interpreting the mechanisms underlying species responses to removal experiments.

Additive and removal experiments are frequently treated as if they were of equal value, the decision being determined solely by convenience. In most cases it is more convenient to remove neighbours than to add them, particularly if they are so large that one cannot raise enough of them to create the desired densities, or if they are so mobile that one cannot keep them confined at the high densities desired. Thus, there are many more removal experiments in the literature than there are additive experiments.

However, more than convenience is involved in the decision because they do not answer the same questions. A removal experiment asks specifically whether reducing *existing* population sizes will provide evidence of competitive release. Thus, it asks whether there is evidence that the community is currently structured by competition. An additive experiment asks whether increasing neighbour abundance above present levels will provide evidence of competition. Thus, even if the answer is a resounding yes, it only shows that competition could potentially structure the community if neighbour densities increased to the level created in the experiment.

Figure 4.6 shows an added complication in the interpretation of additive designs. If one knows in advance the natural maximum population density (vertical line), then one can recognize a stippled zone containing the range of population sizes which could be encountered in the field. (It seems reasonable to assume that the lower limit is always zero.) The additive design may therefore either represent a test for competition in situations which do occur naturally even if they are not now present (left-hand triangle in stippled region), or they may represent densities which never occur in nature but are of theoretical interest (right-hand triangle outside stippled region).

In cases where the perturbation creates population sizes not normally found in nature, additive experiments are more like laboratory experiments except for the much greater realism. As interest grows in predictive ecology, it seems reasonable to expect more use of additive experiments. One can imagine experiments which postulate n different potential future states of a system (a forest with a native herbivore being reintroduced, a marsh downstream from a new hydroelectric dam, a coral reef with new levels of fishing,...), and ask under which possible future states competition will be most (or least) important in controlling species abundances.

Both of the preceding conceptual approaches dealt with testing for pattern and inferring mechanisms. Experiments do not necessarily overcome this problem. Of the three conceptual approaches, the results of experiments eliminate the largest number of alternative hypotheses, but competing hypotheses may remain. As described above, a perturbation really tests for density dependence. One species is reduced in abundance and another increases, but does this demonstrate that competition was occurring between these taxa? The variety of indirect effects which can produce 'apparent competition' have been explored elsewhere (Holt, 1977; Bender et al., 1984; Connell, 1989). Perhaps the removed species was a host for a pathogen which also damaged the remaining species. Perhaps the removed species attracted a predator which also fed upon the species being studied. For example, M. A. Parker and Root (1981) showed that a herbaceous plant species was excluded from some habitats by a grasshopper associated with a common shrub. Springett (1968) showed that *Necrophorus* beetles engaged in exploitation competition by carrying a predator to attack the species with which they were competing (Chapter 1). Such examples suggest that the number of possible mechanisms of interference competition are limitless.

Tilman (1987a) has argued that such competing hypotheses can only be eliminated by explicitly including resource levels in field experiments. For example, in the experiments of Brown et al. (1986) on desert rodents, that resource levels increased in removal plots is strong evidence that competition for resources was occurring. However, this still does not eliminate other alternative hypothses; one could speculate, in view of the above examples of indirect effects, that rodents carry a virus which harms ants, and removing rodents allows resources to increase as well as eliminating the effects of disease.

Perhaps instead of merely monitoring resources, they could be added to some plots, but resource-addition experiments present other problems of interpretation (Chapin *et al.*, 1986).

How are we to determine whether the indirect effects in an experiment were 'apparent' competition (Holt, 1977; Bender *et al.*, 1984; Connell, 1989) or 'real' competition? No change in resource levels need occur if the primary mechanism of interaction is interference. Also, the number of possible mechanisms of interference competition appear to be limitless (for example, Springett, 1968; M. A. Parker and Root, 1981; Boucher *et al.*, 1982; Connell, 1989). Presumably, to distinguish between "apparent" and 'real' competition we must determine whether the indirect effects were coincidental or part of the interference 'tactic' of a competitor. Does the term coincidental have any meaning in this context? How do we decide? If *Necrophorus* beetles carry predators for interference competition (Springett, 1968), then presumably plants can attract herbivores to attack their neighbours (M. A. Parker and Root, 1981). The distinction between 'apparent' and 'real' competition is therefore not clear if one allows for complicated mechanisms of interference competition.

Ultimately the number of competing hypotheses to account for species responses in removal experiments depends upon the latitude of the definition of competition. The broader the range of phenomena termed competition is, the fewer non-competitive hypotheses there are to consider. It may be most useful to recognize that there are many mechanisms of competition, and then to divide them into subcategories such as exploitation competition and interference competition. Interference competition might then be subdivided into direct and indirect. This also emphasizes that one cannot divorce natural history from experiments, since careful observations of mechanism may yield some complex and unexpected methods of interference competition.

As a last comment upon field experiments, it should be noted that experiments in themselves have no inherent virtue unless they are asking a worthwhile question. It is therefore time to consider the three conceptual approaches to the study of competition, and how they can be skilfully used.

4.5 CHOOSING A RESEARCH PATH

There is a danger in comparing and contrasting conceptual approaches. It can leave the impression that the choice of conceptual approach is where research begins. These conceptual approaches are, of course, only paths which offer opportunities for answering questions.

The most important (and probably the most overlooked) part of ecological research is the choice of the question. Important questions either have the potential to falsify significant hypotheses (J. R. Platt, 1964; Loehle, 1987) or to test for the presence of broad general principles (R. H. Peters, 1980a; Stearns, 1982; Leary, 1985). In the absence of generality, ecological research becomes natural history (R. H. Peters, 1980b). Because ecologists usually genuinely enjoy field work, it is all too tempting to rush into data collection without first taking the time to think. Anecdotes about scientists with boxes of data and no idea what to do with them can be found throughout academia. In contrast, once the important question is clearly defined, the obvious route for testing it can emerge naturally. This route has two components: the model system to be used, and the conceptual approach. This chapter has concentrated upon conceptual approaches, but clearly this is not independent of the choice of model system. The two are connected, since different kinds of systems are suitable for answering different kinds of questions. Table 4.4 therefore attempts to put these conceptual approaches into their proper perspective.

Table 4.4 Steps in a research programme which lead naturally to the selection of appropriate tools. For consideration of obstacles at each of these steps, see Chapter 8

---

1. What are the important questions?
     Criteria: generality
                  implications for well-being of other humans and living organisms
2. What specific question will be asked?
     Criteria: generality
                  implications for well-being of other humans and living organisms
                  constraints (time, resources)
                  personal skills
3. What is the appropriate model system?
     Criteria: range of conceptual approaches possible
                  probability of providing clear answer
                  importance in the biosphere
                  possibilities for generality
4. Which data and which tests are appropriate?
     Criteria: operational dependent and independent variables
                  easily measured dependent and independent variables
                  availability of appropriate controls
                  well-defined outcomes with ecological interpretations
                  simplicity
                  elegance
5. Proceed to collect data

Notes for graduate students:
1. Important questions are not important just because others claim they are. Kuhn (1970) observes that scientists tend to choose problems that they think they can solve, not necessarily socially important questions. A little thought may suggest areas where these overlap.
Leary (1985) has provided a framework for evaluating scientific productivity which cross-tabulates generality with type of questions posed.
2. Constraints are to be considered only after the ideal has been set. Do not constrain yourself too early.
3. Note that natural history preferences, convenience and supervisor's preferences are not eligible as criteria.
4. This is where science becomes an art. This is also where the appropriate conceptual approaches emerge naturally. After you have convinced yourself, give several seminars to fault-finding and cynical friends. Repeat steps 1–4. Consult a statistician.
5. By now all the difficult work has been done.

Good model systems allow the broadest range of conceptual approaches. If similar answers emerge using very different conceptual approaches, then we can be far more confident that the results are robust than if only one technique was available. Researchers may find themselves boxed into using only descriptive or comparative evidence for competition precisely because they do not have a clear question and do not plan for generality when choosing their system in the first place. Yet much research is concentrated upon taxa such as birds, which is a relatively insignificant taxon (in numbers of species or biomass) where many kinds of field and laboratory experiments are almost impossible. Ornithologists must therefore often base their arguments upon descriptive studies and comparative studies (Weins, 1983). In contrast, the largest groups of macroscopic organisms in the biosphere are plants and insects (see Figs. 8.2, 8.3) (Colinvaux, 1986) and both provide a remarkable array of life-history types to work with, as well as the possibility of using all conceptual approaches simultaneously.

Price (1984a) also discusses the need to choose carefully the appropriate system for conducting research. His criteria for choosing a study system were as follows: (1) members of the community must be easily defined and recognized; (2) resources must be easily measured; (3) the pool of potential colonists must be clearly defined; and (4) the community and resources must be amenable to extensive experimental manipulation.

Which of the three conceptual approaches is best? By emphasizing the steps in Table 4.4 it is possible partly to sidestep this question. However, certain general statements are possible. Descriptive studies provide the weakest evidence, which is why they probably have provoked the most acrimonious and sterile debates; inferring process from pattern is simply an inefficient research strategy. Comparative studies have more appeal, since they provide the appearance of a 'natural experiment', but they too are dependent upon inferring process from pattern, as Tables 4.2 and 4.3 illustrate. It is easy to assume that demonstrating a statistically significant pattern demonstrates the existence of competition. These are not equivalent, and it seems that competing hypotheses to account for the patterns observed are rarely discussed. It therefore appears that comparative studies have most of the weaknesses of descriptive studies. A possible exception occurs when a comparative study is set up to answer a specific question. If the question comes first, then a comparative study is far more valid than if data are collected and then used to tell a story consistent with the observer's beliefs. Field experiments are powerful because they usually (but not always) require a clear initial question and have proper controls. When badly designed, as they often are (Hurlbert, 1984), they may be as uninterpretable as any other data. Recently published studies in major ecological journals suggest that Hurlbert's advice is still being overlooked, and Salt (1983) has observed: 'because of faulty or inappropriate design and execution, a great many experiments prove nothing except that the individual attempted an experiment'. Another problem is that the mechanisms which produce species responses to removal experiments may be both complicated and obscure. A popular criticism of field experiments is that they lack generality, which was one of the two principal criteria for an important question (Table 4.4). However, this is not an inherent trait of field experiments, as the next chapter shows. Hybrids of comparative studies and field experiments are possible, and are explored further in Chapters 5 and 7.

A potential weakness of all three approaches is that in the absence of clear, well-justified questions they provide a cloak under which it is possible to justify: (1) free-form data collection; and (2) selection of systems for natural history preferences rather than for sound scientific reasons. However, it is obvious that descriptive studies are most likely to have this flaw, whereas experimental studies, which force users to plan a design, are least susceptible. By emphasizing the question first and the choice of system and conceptual approach second, many potential problems of interpretation can be avoided. This has the advantage that more time is left for concentrating upon novel approaches to answering the questions being posed. Once the questions are well defined and there is time to reflect, then the tools become obvious.

Perhaps ecological societies will one day host conferences where no data are permitted. At such conferences researchers will justify their question, their model system and their research strategy. Editors could also help by insisting that introductions to papers present clear questions and justification for the choice of model system and research strategy. Until then, the bottom line might be paraphrased as follows: Caution–engage brain before collecting data.

The next three chapters will explore some new research paths that search for general principles regarding competition. In Chapter 5 the potential for expanding the generality of field experiments is examined – the conclusion is that options to increase generality have barely been explored, and that many possible innovative designs are possible. In Chapter 6 laboratory experiments are investigated and the significance of competitive hierarchies in communities is considered. Chaper 7 explores communities where competition may be unimportant, and discusses a paradox about the relationship between resource levels and competition intensity. Chapter 8 returns to the steps in Table 4.4 to explore the future prospects for competition theory.

4.6 QUESTIONS FOR DISCUSSION

1. What are the strengths and weaknesses of the descriptive, comparative and experimental approaches?

2. Why have ecologists spent so much time inferring process from pattern? Do we have any methods of obtaining direct information on process?

3. Plant ecologists have devoted much time to discussing pattern analysis, developing techniques and debating their interpretation. What, if anything,

was learned during this process? Why have zoologists resurrected the debate? Why have they done it without reference to the lessons of plant ecology?

4. 'Ecology would advance more rapidly if for 2 years we declared a moratorium on data collection, and instead concentrated on critical reviews of questions and methods.' Do you agree?

5. 'The choice of model systems is one of the least-thought-about steps in ecological research.' Do you agree? Why should this be the case?

6. Consider the regions of enquiry in Fig. 4.6. Does one of these areas deserve scientific priority? If so, why?

7. '... the hand that pulls the lever on the voting machine also releases the bombs on the helpless villagers in the bombadier's path' (Friedenberg, 1976). Discuss the implications of this quotation for unravelling the mechanisms of competition.

5 Extending the generality of field experiments

General notions are generally wrong.

M. W. Montagu, 1710

Caeser had his Brutus – Charles the First, his
Cromwell – and George the Third...may profit by
their example.

Patrick Henry, 1765

To do science is to search for repeated patterns, not
simply to accumulate facts....

R. H. MacArthur (1972)

An objective of competition theory and, in fact, of any ecological theory, has to be the ability to make general predictive statements about the structure and behaviour of ecological communities. We have already explored the difficulties in arriving at general conclusions through collecting special cases (Chapter 2), and R. H. Peters (1980b) has provided a useful reminder that in the absence of general principles ecology becomes natural history. Therefore, Table 4.4 listed generality as one of the most important criteria to consider when planning ecological research.

The conceptual approaches reviewed in Chapter 4 differ in the generality they provide (Table 4.1; see also Diamond, 1986), with experiments allegedly yielding the least generality. Yet field experiments also offer some important advantages. Are there inescapable trade-offs between descriptive, comparative and experimental approaches, or is a reconciliation possible? The objective of this chapter is to review some criticisms of field experiments, and to argue that, rather than being inherent limitations of field experiments, the problems lie with the way in which they have so far been employed. By starting with general questions and planning the experiment so that generality is maximized, it is possible to make field experiments much more powerful. In this way we can have generality and the rigour associated with a *priori* questions and experimental design. Five principles for increasing generality are explored. These are:

1. examining general patterns in the first place
2. using increased numbers of species
3. using comparative ecology, rather than species, nomenclature
4. using general experimental factors
5. arranging treatments along natural gradients.

5.1 CRITICISMS REGARDING LACK OF GENERALITY

The problem with lack of generality in field experiments was raised in Chapter 2, as well as by Diamond (1986) and Tilman (1987a). Although the criticism – lack of generality – is the same, these three sources provide fundamentally different reasons for making this criticism.

Diamond (1986) considers lack of generality to be an inherent trait of field experiments because they are by necessity limited in time and space. Diamond's view is that field experiments lie midway along a continuum from laboratory experiments (high in precision and low in generality) to comparative studies ('natural experiments'; low in precision and high in generality). This trade-off therefore produces inherent weaknesses in field experiments. In fact, they are unsatisfactory because they lack both the precision of laboratory experiments and the generality of comparative studies! The implication is that this inherent weakness cannot be overcome, although Diamond (1986) does list ten possible modifications to increase the value of such experiments.

Tilman (1987a) does not criticize field experiments *per se*, merely the way in which they have been conducted. He argues that the lack of generality is caused by the lack of emphasis upon mechanism (see Chapter 4). In the absence of consideration of mechanism, field experiments simply describe phenomena and contribute to a sea of observations whose interpretation is difficult and from which the drawing of generalizations is even more difficult. The problem of interpretation of competition experiments lies in the difficulty of interpreting removal experiments; there are many possible indirect effects which could cause one species to increase when another is removed (Holt, 1977; Bender *et al.*, 1984; Connell, 1989). Therefore, removal experiments by themselves do not demonstrate competition, Tilman argues; only experiments which manipulate resource levels can do this. Even if there is further evidence demonstrating that competition probably occurred, the absence of mechanism prevents the comparison of different studies and therefore prevents the drawing of general conclusions.

Chapter 2 also criticizes field experiments for being phenomenological, but for a different reason. This criticism was not based upon inherent traits of field experiments, nor upon the weakness of removal experiments, but simply on the lack of planning for and explicit consideration of generality. Sites appear to be chosen for convenience or natural appeal. Organisms appear to be chosen because they are popular or attractive, rather than because they have the potential for revealing important general principles. The choice of species and sites therefore often appears to be completely arbitrary, and this limits attempts to draw general conclusions about the biosphere.

Each of these viewpoints is in turn open to criticism. By placing emphasis upon the inherent limitations of field experiments, Diamond implies that there is little which can be done to address the limitations of field experiments, which may encourage continued emphasis upon descriptive and comparative studies. Tilman (1987a) can be criticized on three counts. The first two are discussed in Chapter 4, and deal with the interpretation of field removal experiments. Briefly, experiments which supplement resource levels may not, in fact, yield unambiguous conclusions about mechanisms (Chapin *et al.*, 1986). Secondly, interference competition does not change resource levels, only access to them, and there are many possible indirect forms of interference competition. The third problem is that the simple accumulation of large numbers of cases where mechanisms are studied does not guarantee that generalizations will emerge. The collection of examples of mechanisms may itself be phenomenological. Finally, the points raised in Chapter 2 are also open to criticism; carefully designing for generality may reveal general patterns of response to removals, but responses to removals may be difficult to interpret (Chapter 4). Also, some might argue that the choice of organisms and systems is not part of the scientific process; perhaps choice of organism is a given, much like a scientist's personality, and expecting scientists to choose organisms because of their importance in the biosphere or suitability for answering questions is simply psychologically unrealistic (we return to this in Chapter 8).

Whichever of these three viewpoints is considered to be important, there is no excuse for simple sloppiness and carelessness. Hurlbert (1984) observed that poorly designed or incorrectly analysed experimental work is flooding the literature. Moreover, 'statistical sophistication or the lack of it is not the main problem. At least in field ecology, the designs of most experiments are simple and when errors are made they are of the gross sort'. It is obvious that incorrectly designed and analysed experiments will contribute little to the advancement of competition theory, or of science in general.

Having explored these criticisms of field experiments, let us turn to resolving them. Although field experiments do have inherent limitations, they have not yet been pushed to their limits. Telescopes also have their limitations, but in terms of designing field experiments, ecologists are still at very early stages in designing optical systems and producing large mirrors. The following principles require practitioners to state clearly the goals of their experiment in advance, and then to consider carefully the degree to which generality can be enhanced at the stages of site selection, species selection and experimental design. Each principle is briefly discussed and then illustrated with examples from the recent literature.

5.2 DEMONSTRATING GENERALITY OF PATTERN

A field experiment is set up to investigate a pattern. Clearly, if the pattern is widespread, then the results are of more interest than if the pattern is a special case found nowhere else on the planet. At its simplest, the person designing the experiment can use intuition and choose a system which most reasonable observers will agree is typical and widespread. For example, the studies of zonation in *Typha* marshes (Grace and Wetzel, 1981) and salt marshes (Snow and Vince, 1984) probably fall into this category. However, more-sophisticated approaches are possible. The extensive literature on detecting pattern in ecological communities referred to in Chapter 4 can be regarded as a useful tool at this stage of experimental design. Weins (1981) has shown that single sample surveys of communities can produce misleading interpretations, which emphasizes the need to demonstrate the generality of a pattern before experiments are begun.

In studying the exposure gradient on lakeshores on the Canadian shield, S. D. Wilson and Keddy (1985) specifically tested whether plant distributions along this gradient were concordant among four different lakes (Fig. 5.1).

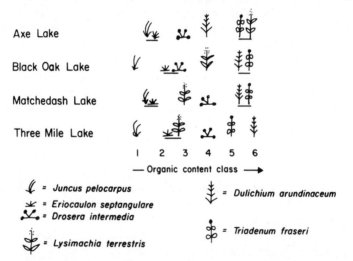

Figure 5.1   The distribution of six lakeshore species along an organic content gradient in four different lakes on the Precambrian shield in Ontario, Canada. In addition to relying on field observations from an extensive sample of lakes, S. D. Wilson and Keddy (1985) used this sub-sample of four to specifically test whether there was a pattern before beginning their field experiments. They used Kendall's test (Siegel, 1956) to test for rank concordance of species distributions along a gradient of six organic content classes and found the ranking was similar among lakes ($P < 0.01$); ties are indicated by underlining (data from S. D. Wilson, 1986). An analysis like this also aids in choosing species for experimental manipulation. In this case *Juncus pelocarpus* and *Dulichium arundinaceum* have particularly consistent patterns, whereas *Lysimachia terrestris* shows considerable among-lake variation.

Table 5.1 Demonstrating generality of pattern before beginning a study. The relative abundances of chalk grassland species in nine different published studies (after Mitchley and Grubb, 1986)

	Literature source									
Species	1	2	3	4	5	6	7	8	9	*Mean rank*
Sanguisorba minor	2	1	4	2	1	2	3	1	1	1.9
Leontodon hispidus	7	2	3	6	4	4	6	9	4	5.0
Lotus corniculatus	5	5	1	4	5	6	7	6	10	5.4
Cirsium acaule	6	3	5	1	3	5	14	5	3	5.6
Plantago lanceolata	1	4	2	7	8	8	9	2	11	5.9
Helanthemum nummularium	12	9	11	5	11	1	2	3	2	6.9
Hieracium pilosella	15	14	17	8	2	9	4	11	5	9.4
Hippocrepis comosa	11	11	12	9	6	11	1	13	6	9.9
Succisa pratensis	4	6	6	12	21	12	24	7	19	10.9
Thymus praecox	19	19	25	3	10	3	5	4	11	11.1
Scabiosa columbaria	10	12	10	10	13	14	9	18	8	11.7
Plantago media	3	8	7	13	15	15	18	15	14	12.1
Centaurea nigra	14	22	21	15	17	10	13	8	7	12.8
Filipendula vulgaris	21	16	15	10	7	19	12	13	13	14.1
Campanula rotundifolia	16	18	18	18	14	13	8	17	9	14.6
Pimpinella saxifraga	17	15	13	16	20	15	9	20	15	15.8
Prunella vulgaris	9	13	9	21	19	22	20	12	18	16.0
Asperula cynanchica	13	10	20	23	17	7	19	22	20	16.8
Viola hirta	25	17	16	14	23	18	16	10	17	17.4
Ranunculus bulbosus	18	21	18	19	12	24	14	24	16	18.6
Trifolium pratense	8	7	8	24	25			25	23	20.1
Campanula glomerata	23	20	22	19	21	17		21	21	20.1
Primula veris	20	23	14	17	24	25	20	16	23	20.3
Polygala vulgaris	22	24	24	21	16	20	16	19	22	20.6
Achillea millefolium	24			25	9	20	20	23	25	22.1
Astragalus danicus				26		23	20	26	26	24.4
Phyteuma orbiculare			23	27	26					25.3

Similarly, in a study of competition among chalk grassland plants, Mitchley and Grubb (1986) compared data from nine published studies to test whether the observed pattern of species abundances was site-specific or general (Table 5.1). Procedures such as this could be required as a routine first step in the design of a field experiment.

5.3 USING INCREASED NUMBERS OF SPECIES

Most experimental studies are based upon comparisons of responses in a few selected taxa. How dependent the results are upon the particular species chosen is not known. This problem could be overcome by including as many species as is practicable within the experiment. Considering the time and effort involved in designing a convincing experiment, the increased number of species may be a relatively small extra cost. The trade-off may ultimately be whether to pose complex questions with a few species, or simple questions with many. Perhaps the criticism of lack of generality arises in part from the tendency of field experimentalists to opt for the former rather than the latter.

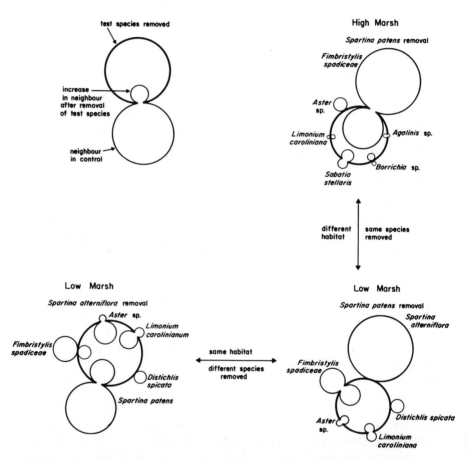

Figure 5.2   A competitive release experiment in a multispecies community of coastal plants. The panel on the upper left shows how to interpret the remaining figures; circle size is a measure of abundance either before or after the experimental manipulations. The right-hand figures compare the effects of removing the same species from two habitats (removing *Spartina patens* from high and low marsh); the bottom figures illustrate the effects of removing two different species from the same habitat (*Spartina alterniflora* and *S. patens* removed from low marsh). Only the response of *Fimbrystylis* is statistically significant. (After Silander and Antonovics, 1982.)

Consider two examples where large numbers of species were included in removal experiments.

Silander and Antonovics (1982) used pairwise removals to explore competitive interactions in different plant communities along an environmental gradient from dunes to low marshes along the coast of North Carolina. Experimental removals were carried out in five vegetation types, and in each case the response of neighbours to removal of a test species was measured by comparing cover in treatment plots with cover in unmanipulated controls. The design allowed measurement of specific as opposed to diffuse responses as well as asymmetry. Figure 5.2 shows some typical results, illustrating a case where different species were removed from the same habitat, and where one species, *Spartina patens*, was removed from two habitats. In a second example, Keddy (1989b) conducted a 4-year experiment on the shoreline of a small lake. The objective was to test whether shrubs controlled the upper distributional limits of herbaceous species on shorelines. Twenty-five plots were paired with controls, then all shrubs were removed from the treatments. The colonization of all other shoreline plants in the cleared and control plots was then monitored over a 4-year period. More than 50 naturally occurring shoreline plants in the adjacent vegetation could have invaded the clearings, but at the conclusion of this experiment the species on the lakeshore fell into two categories: one subset had invaded the cleared plots, suggesting that their upper limits were set by shrubs, whereas the other group showed no significant differences between treatment and control plots.

These two removal experiments differed somewhat in their approach, but both monitored the responses of a large number of species. They also illustrate the difficulties both in analysing and in interpreting such experiments. The very large number of species and sites produces a large number of possible pairwise comparisons. Listing the names of species which responded to specific treatments means little to those who are not specifically concerned with that particular model system. Thus, there are major constraints upon the drawing of conclusions which might apply to other systems. This problem leads naturally to a consideration of comparative ecology, which provides a tool for escaping from the restrictions imposed by species nomenclature.

5.4 PROVIDING A COMPARATIVE CONTEXT

An alternative to including many species in an experiment is to select them because they represent a larger group of species about which one wishes to draw inferences. It can be made far more systematic by specifically defining functional groups ('guilds', 'strategies' or 'life-history types') in advance of the experiment, and choosing species to represent these groups. Studies of resource partitioning may provide a guide to comparing the effects of competition within and among different guilds. In some systems preliminary classifications of life-history types or guilds may already be available.

Cummins (1973) and Cummins and Klug (1979) have recognized different feeding groups of aquatic invertebrates. Severinghaus (1981) has proposed a preliminary classification of major groups of terrestrial animals in the USA. There is a long history of defining functional groups based upon plant morphology (du Rietz, 1931), and other life-history traits can also be included (for example, Grime, 1974; Grubb, 1985). In other systems preliminary screening may be necessary. The techniques of comparative ecology pioneered by Grime and Hunt (1975) and Grime *et al.* (1981) offer many opportunities for systematically screening species for recognizing such functional groups; the state of the art is reviewed in a recent symposium volume (Rorison *et al.*, 1987). As an example of combining screening with field experiments, Shipley (1987) conducted a series of transplant experiments along freshwater shorelines to test among models for plant zonation; the generality of the results of his experiments was increased by screening a large number of marsh species for life-history, morphological and physiological traits. The experimental species were included in this screening so that the degree of similarity between the test species and other members of the marsh flora could be assessed quantitatively.

Although screening in advance of the experiment would probably be the best approach, it can also be done after the analysis in order to try to generalize the significance of the results. It was noted above that giving the names of species responding to removals really does not permit many general conclusions to be drawn from studies. A difference between the two studies is that Keddy (1989b) attempted to increase the generality of conclusions by screening species for life-history traits and growth rates; groups responding and not responding to the treatment were found to have basic differences in life history and growth rates (Table 5.2). It appears that smaller species with

Table 5.2 Using traits rather than species' names to increase the generality of results. The traits of seven species of lakeshore plants which increased after removal of dominant shrubs contrasted with the traits of six species which showed no response (after Keddy, 1989b)

Trait	Response to removal experiment[a]		
	Increased	No change	
Mean shoot height (cm)	6.7	27.0	***
Mean dry weight (g)	0.02	0.33	*
Growth rate (g g^{-1} day^{-1})	0.20	0.07	*
Density of buried seeds (m^{-2})	44.0	1.0	***
Seed weight ($\times 10^{-3}$ g)	0.12	0.90	*
Habitat			
Mean organic content	6.4	11.9	**
Mean elevation	16.4	20.8	NS
Abundance (% cover)	22.3	22.7	NS

[a] Mann–Whitney U-test, two-tailed. * $P < 0.05$; ** $P < 0.01$; *** $P < 0.001$.

buried seed reserves colonized cleared plots more successfully than did large species with vegetative propagation. This could be interpreted to mean that large plants were not inhibited by shrubs because they could grow through the canopy and avoid shading, or that the experiments did not run long enough, so that only natural gap-colonizers had time to respond to the perturbation. It is clear that Table 5.2 is of much greater general use than a list of the species names in the two categories. Another example of this appears in the removal experiments done by Dayton (1975), discussed in Chapter 7, where, species responding to removals were divided into canopy and fugitive species (Fig. 7.2).

5.5 USING GENERAL EXPERIMENTAL FACTORS

The factor or factors manipulated in the experiment should be chosen to have the broadest possible generality. An example of this would be the manipulation of biomass in plant communities. One strong generalization in plant community ecology is that plant species richness increases and then decreases with increasing plant biomass (for example, Al-Mufti et al., 1977; Grime, 1973, 1979; Silvertown, 1980; Tilman 1982, pp. 123–32; Austin, 1986; Day et al., 1988; Moore et al., 1989).

Such generalizations are possible only because different researchers concentrated upon similar dependent and independent variables: species richness and biomass. The importance of such state variables to general theory has been discussed elsewhere (for example, Lewontin, 1974; Rigler, 1982; Keddy, 1987). Biomass is a function of soil nutrients, so relationships between species richness, biomass and soil fertility can be compared across very different plant communities.

The Park Grass experiments at Rothamstead, UK, are frequently cited because they represent an experiment which has run for more than a century. What makes them of greatest interest, however, is that they deal with dependent and independent variables of interest to many ecologists: species richness, biomass and soil fertility. Silvertown (1980) and Tilman (1982) have therefore carried out extensive analyses of patterns in these vegetation attributes. Figure 5.3 shows that the relationship between species richness and biomass is remarkably consistent. The top line represents the situation more than 80 years before the two bottom ones, yet the slopes are identical. The bottom lines both represent the effects of fertilization. Fertilization reduced richness, but those plots severely acidified by addition of ammonium sulphate (open circles) maintained an identical slope. Tilman (1982) adds several other generalizations: (1) the rate of competitive exclusion is slow – sometimes the full effects took a century to occur; (2) complete fertilization increased biomass and led to dominance by one or two species – in all such cases plant growth was probably limited by light; and (3) which dominant occurred in a plot depended upon the pH of the soil.

Studies which experimentally manipulate these variables or use biomass gradients as a predictor variable will be of considerable general interest in

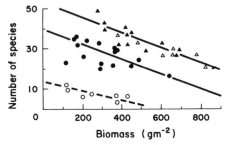

Figure 5.3 The relationship between species richness in experimental plots in the Park Grass experiments at Rothamstead, UK. The plots in the top line (triangles) represent the situation before fertilization; the bottom two lines illustrate the effects of fertilization. The open circles represent plots which were acidified when fertilized. The top line precedes the bottom lines by 80 years. In all three cases, across time and treatments, the slopes are nearly identical (after Silvertown, 1980). What makes this study of particular interest is the use of dependent and independent variables which can be easily generalized to other situations.

developing and testing ecological theories (Keddy, 1987, 1989a).

Even if a question is posed about a specific ecosystem, and an experiment is carried out under artificial conditions, the design can be modified to maximize generality. When S. D. Wilson and Keddy (1985) posed a question about the physiological responses of plants to the shoreline gradient illustrated in Fig. 5.1, they collected soil from five different lakes and combined it in pots to produce a generalized shoreline substrate gradient. This approach is obviously applicable to many other natural gradients.

5.6 ARRANGEMENT ALONG GRADIENTS

Experiments are often set up at only one location. Obviously, the larger the experiment is, the greater the generality. There is no reason why experiments could not be designed with replicates scattered over a much larger area. One could imagine a study of prairies with different blocks of the experiment in five different prairie relicts scattered across the North American midwest. A more cost-effective alternative may be to exploit naturally occurring environmental gradients and to set up experiments along these gradients. Not only is spatial variation then incorporated but, because the extra variation is systematic, it is possible to ask how the treatment effects vary along this gradient. The use of naturally occurring gradients is a well-established tradition in experimental ecology. The rocky intertidal zone offers a rich literature of studies where both distribution and abundance have been measured after experimental removals (Connel, 1961, 1972; Lubchenco, 1980), although many of these are inadequately replicated or pseudoreplicated (Hurlbert, 1984). Gradients in salt marshes (Snow and Vince, 1984), grasslands (Gurevitch, 1986) and freshwater shorelines (Grace and Wetzel, 1981; Shipley, 1987; Keddy, 1989b) have all been used in removal experiments.

A removal experiment along an environmental gradient is probably the best of all possible field experiments for studying competition. Such experiments answer not only questions about factors controlling the abundance of species, but also questions concerning their distribution. Moreover, these sorts of experiments provide a means of measuring and testing hypotheses concerning realized and fundamental niches. Pielou (1977) drew attention to this a decade ago, observing that gradients '... are far more deserving of study than the elusive "homogeneous environments" traditionally beloved of ecologists'.

A classic study by Connell (1961) used experimental removals along an environmental gradient to test whether competition controls the distribution of intertidal barnacles. Connell noticed that two species of barnacles occupied separate horizontal zones on rocky coasts. Although the adults were spatially separated, the young of the upper species, *Chthamalus stellatus* settled but did not survive in the lower zone occupied by *Balanus balanoides*. Connell postulated that the upper species failed to occur lower on the shore because of competition from the lower species. In one series of experiments he removed *Balanus* from plots paired with controls where *Balanus* was present, and monitored survival for 1 year. The dependent variable was survival, and Fig. 5.4 shows survival in treatment and control plots at two heights on the shore. On the upper shore (left) removal of *Balanus* had little effect upon the *Chthamalus*, whereas on the lower shore (right), removal of *Balanus* allowed the *Chthamalus* to survive. Moreover, the *Chthamalus* were actually more

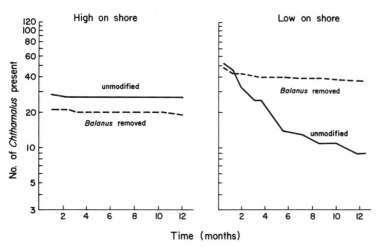

Figure 5.4   A competitive release experiment using animals which occupy an environmental gradient. The survival of the barnacle *Chthamalus* is shown in plots with and without crowding from *Balanus* for plots higher on the shore (in the *Chthamalus* zone) and lower on the shore (in the *Balanus* zone).

abundant on the cleared plots on the lower shore than they were on the upper shore where they normally occur. These results provide strong evidence that the upper species, *Chthamalus*, is competitively displaced from the lower shore. Connell supplements this experiment with many observations and details of natural history consistent with this interpretation.

Similarly, Dayton (1975) set up removal experiments of macro-algae along an exposure gradient in the rocky intertidal zone, and was able to show that the effects of the removal varied among sites; this experiment is discussed in more detail in Chapter 7. The lakeshore field experiment described above (Keddy, 1989b) also included this approach. The 25 paired plots were located along an exposure gradient. In addition to asking whether there was evidence of competitive release, it was possible to ask how the results depended upon position along the naturally occurring stress–disturbance gradient. Although the experiment of Silander and Antonovics in Fig. 5.2 also compares several habitats, interpretation is more difficult because we do not know enough about the habitats that they used to generalize to other types of gradients.

Such designs using natural gradients combine the rigour of planned and controlled manipulations with the generality of among-site comparisons. At this point field experiments begin to take on attributes of comparative studies (Chapter 4). In fact, this combination of natural gradients and experiments may be the most powerful tool available for probing the structure of ecological communities. Chapter 7 takes up this theme by exploring current ideas about where different intensities of competition are likely to be found in nature.

The scope for these sorts of studies could be greatly increased if ecologists co-operated more. If similar experiments were repeated simultaneously in different geographical regions, there would be a much better opportunity to assess whether the observed responses were site specific, or whether they represented broad general principles. J. H. Connell (pers. commun.) reports that a co-operative study is under way comparing competition and zonation on two different continents. Perhaps more co-operation and less competition among ecologists could lead to similar ventures.

5.7 CONCLUSION

There are five principles for expanding the generality of field experiments. The pattern explored must be demonstrably general. More species can be included. Life-history traits and functional groups can replace species nomenclature. General environmental factors can be manipulated. Natural environmental gradients can be built into the design. These five principles provide five clear criteria for judging the potential of a field experiment to produce results of general significance. Experiments which combine two or more of these principles offer a route for innovative and exciting experimental designs. Chapter 4 concluded that questions should be posed in advance of data collection. Generality requires the same consideration.

5.8   QUESTIONS FOR DISCUSSION

1. What are the three criticisms of field experiments which were presented? Are they valid?

2. Why have so few studies failed to plan for generality of results and conclusions?

3. Read the 'Methods' sections of recent papers on competition. Are they intelligible? Is the design properly described, and are the data properly analysed?

4. Consider each of the five principles for expanding generality. Are there better published examples?

5. The examples cited in this chapter deal with sessile organisms. What hypotheses could account for this pattern, and which is the most plausible?

6. Why do ecologists have a fascination with homogeneous systems?

7. Why are there so few collaborative studies across large temporal and spatial scales?

6 Community matrices and competitive hierarchies

All animals are equal, but some animals are more equal than others.

G. Orwell (1945)

There is little friendship in the world, and least of all between equals.

F. Bacon, Essays

How does competition produce organization in communities? Although Chapters 2 and 4 show that competition does control the abundance of species in some communities, to derive general principles we need systematic studies of the way in which entire communities are organized by competition. Here is where field experiments are at their weakest. In natural communities we are presented with mixtures of species in varying abundances, at different developmental stages and with poorly known histories. Rare members of the community will be particularly difficult to study, since there may not even be sufficiently large populations to design proper experiments. This is where laboratory experiments provide a powerful tool. We can raise individuals of all populations that we wish to study, allow them to interact in pairs, threes and so on, and actually map the linkages produced by competition. This map could represent the potential structure upon which all other agents of community organization act. By providing different experimental environments, it would then be possible to test whether the map is deformed by environments in predictable ways.

Do such maps exist? Here there is an interesting gap in the data. There are many studies which have constructed community matrices and explored their properties (for example, Vandermeer, 1970; May, 1974; Levins, 1975; Yodzis, 1978), but a majority of community matrices are based upon descriptive data, or measures of 'niche overlap', which do not really measure competition at all (Chapter 3). On the other hand, laboratory experiments frequently examine only a few pairs of species (Gause, 1932; Park, 1948, 1954; Longstaff, 1976; Widden, 1984) so it is difficult to picture how these few pairwise interactions fit into a larger whole. Exceptions exist. Figure 6.1 shows the pairwise interactions in an experimental fly community (right) and in two experimental plant

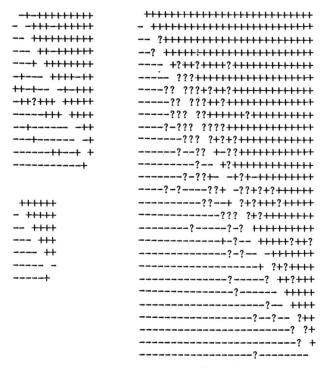

Figure 6.1 Three published community matrices based upon experiments. Plus signs and minus signs indicate whether the species represented by each row was dominant over or suppressed by the column species. In order from largest to smallest the matrices are from *Drosphila* (Gilpin *et al.*, 1986), sea-cliff plants (Goldsmith, 1978) and lakeshore plants (S. D. Wilson and Keddy, 1986b). Other examples can be found in Buss (1979), Fowler (1981) and Keddy and Shipley (1989).

communities (left). Could you tell which was which without the captions? This chapter first explores the way in which such matrices can be experimentally derived and the techniques available for detecting pattern. Some existing examples are then described. It appears that many communities have such competitive hierarchies. The consequences of this for competition theory and community organization have been little explored. Those interested only in the results and consequences need not read the following methodological section on detecting patterns in community matrices.

6.1   PATTERNS IN COMMUNITY MATRICES

The paucity of published community matrices makes it difficult to draw generalizations about patterns or mechanisms in such communities. The above matrices suggest that both plant and insect experimental communities have competitive hierarchies. That is to say, species can be ranked in order of competitive ability from a competitive dominant which is able to suppress all other species to a subordinant which is suppressed by all other species. This section provides a brief introduction to the construction of such matrices and the search for patterns in them.

6.1.1 Constructing experimental matrices

To produce matrices such as those in Fig. 6.1 requires the measurement of many interspecific competitive abilities simultaneously. Two steps need consideration. The first is the design of an experiment to assess the many pairwise interactions. The second is the analysis and interpretation of the results. Consider these in turn.

The design of experiments to detect competition in laboratories provides experimentalists with several decisions. In any such experiment it is essential to have control populations grown by themselves. Unless such monocultures exist and persist under the experimental conditions provided, the disappearance of species from mixtures may simply mean that the physical conditions were unsuitable for some of the species concerned. The first decision regards the way in which these monocultures are used as references for mixtures; i.e. whether to use additive or substitutive design. In an additive design the mixtures have twice the density of monocultures; in substitutive, the mixtures have the same density because each experimental population is inoculated in the mixture at half the density of monocultures. The latter replacement series design is discussed further in de Wit (1960), Harper (1977) and Firbank and Watkinson (1985). The principal strength of replacement series designs is that performance in monoculture and mixture is measured at the same density. This is also a weakness; this concern with constant density is only reasonable if all species are of similar size, so that equivalence of density translates into equivalence of biomass. Recall the problem of comparing ants and rodents (Chapter 2). If experimental matrices compare organims of different sizes (and they must do so to explore the structure of most real communities), then this concern with equivalence of density seems unimportant. A second strength of the replacement series design is that it enables comparison of inter- with intraspecific competition – indeed, this is the theoretical context in which such studies are normally carried out (Harper, 1977). However, in constructing community matrices and testing for competitive hierarchies, the primary goal is to measure as many interspecific interactions as possible. If each species' interspecific competitive ability is scaled relative to its intraspecific competitive ability, then comparison of interspecific competitive ability is confounded. Since intraspecific competitive ability varies among species, the interspecific competitive ability of each species is being measured on a different scale. Additive experiments may therefore be the simplest to interpret. Since most published plant examples use replacement series designs, however, the following sections are largely based upon these in spite of their shortcomings.

Part of the difficulty in interpreting such experiments results from the confusion of two experimental effects. The total effect of one species on another is the product of: (1) the effect per unit biomass; and (2) the total biomass. Experiments which hold biomass constant compare only the effect per unit biomass of two species. It is not that this is wrong, but if (say) the experiment compares ants with rodents, or shrubs with herbs, then the relative impact of the two species upon each other is almost certainly a consequence of the differences in biomass. That is, even if one species is superior to the other on per-biomass effects, if the other species is always much more abundant because of inherent differences in size, the competitive dominant may well be the species with lesser effects per unit biomass. In any analysis with pairs of very similar species in culture, only the per-biomass effects are being measured. The design has specifically excluded the second effect, even though it may be very important in real communities. Any design, analysis or interpretation of competition experiments needs to keep these two effects separate. The tendency to think in terms of Lotka—Volterra equations which consider population size as a number N has perhaps contributed to the confusion, since it hides consideration of size-dependent effects. Similarly, the constraint in replacement series experiments that individuals must be of similar sizes has probably delayed experimental assessment of biomass effects.

Another unresolved question is whether and how to renew resource supplies, since renewing resource supplies can decrease the degree to which resources limit growth. The relative performance of two competitors under moderately high resource levels may be quite different from that when resources are low and the food is contaminated by waste products. If contamination by metabolic wastes is a mechanism of interference competition, resource renewal may prevent certain competitive mechanisms from occurring. Changes of media also frequently lead to some mortality, whether it be algae in chemostats or flies in bottles; these must be carefully considered in interpreting outcomes. In vascular plant experiments, increasing nutrient levels probably shifts the interaction from competition for nutrients to competition for light. In contrast, for algae in chemostats, increased nutrient levels may simply change the relative growth rates of the competing species.

Interpreting the outcome of these experiments also requires certain decisions. What exactly is meant by the plus and minus signs in Fig. 6.1? In the limiting cases a minus could mean that a species was extirpated in mixture, and a plus could mean that only the other species persisted. In practice such clear results may not occur, if only because laboratory experiments comparing this number of species may not run for sufficiently long times. The winner in a pairwise interaction may therefore be defined as the species with superior performance after a certain period. Superior performance is usually defined as greater biomass or larger populations in mixture. This may or may not be standardized relative to control performance. In some cases exact criteria are not given. Gilpin *et al.* (1986) merely state that experiments were carried out 'until the conclusion was obvious'.

Recall that the objective of such experiments is to compare the interspecific competitive abilities of as many species as possible. Few experiments have set this goal, so the design and interpretation of such experiments has not yet been standardized in any way. Interpretations of the existing literature must therefore frequently use data which are less than ideal. If experiments are designed specifically for exploring some of the following questions, then the situation will be improved.

6.1.2   Competitive hierarchies

Once the matrix is available for inspection, one obvious pattern to look for is competitive hierarchies. It appears that some species are able to suppress all other species that they encounter, whereas other species are suppressed by all species that they encounter. Species could then be ranked in order from competitive dominant to competitive subordinant. The simplest measure of position in the hierarchy may be the number of proportion of species suppressed (e.g. Fig. 6.2; S. D. Wilson and Keddy, 1986b), perhaps weighted by the degree of suppression. McGilchrist and Trenbath (1971) have provided methods for calculating a measure of aggressivity for each species in

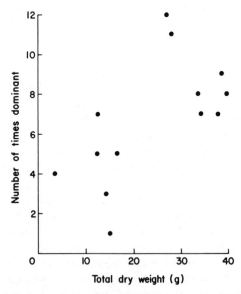

Figure 6.2   The relationship between position in the competitive hierarchy (measured as number of times dominant) plotted against plant size in monoculture for sea-cliff plants (data from Goldsmsith, 1978). Large plants suppress small ones, but the relationship is less clear for plants that are very similar in size ($r = 0.61$; $P < 0.05$).

replacement series experiments (although the interpretation of this measure is difficult when relative yield totals are greater than unity: i.e., when there is resource partitioning as well as competition; L. Aarssen, pers, commun.). The presence of a hierarchy may be obvious upon inspection (e.g. Fig. 6.1), but this may not always be the case. Keddy and Shipley (1989) have therefore proposed a simple statistical test for the presence of hierarchical organization in such matrices. This test converts the matrix to binary form and tests for deviations from a null model.

6.1.3   Asymmetry

Competitive hierarchies should be associated with asymmetric competition. That is, if there is competitive hierarchies, then when two species are put into mixture one would be greatly suppressed and the other little affected. A simple measure of asymmetric competition would be to measure for each of two species the degree to which it is suppressed by the presence of the other. If both species are equally suppressed, then the interaction is symmetric. The greater the difference in responses of the two species is, the greater the asymmetry. By determining pairwise interactions experimentally, an accurate measure of competitive asymmetry can be produced for each pair of species. It is analogous to the measure of reciprocity of Silander and Antonovics (1982) in field experiments.

For example, in plant experiments results of replacement series experiments are frequently reported as a matrix of values of relative yield per plant (RYP). If Y_{ii} is defined as the yield of an individual of i in a monoculture and Y_{ij} as the yield of an individual of i in mixture with j, then $RYP_{ij} = Y_{ij}/Y_{ii}$: its expected value in the absence of competition is $RYP_{ij} = RYP_{ji} = 1$. A simple measure of asymmetry would therefore be

$$A = |(RYP_{ij} - 1) - (RYP_{ji} - 1)|$$

Asymmetry will therefore usually be largest when one value is greater than unity and the other less than unity. However, if both species have $RYP > 1$, or $RYP < 1$, then A measures the degree to which they differ in their responses to growing in a mixture. A value can therefore be calculated for all pairs of species reported in published diallele tables.

Digby and Kempton (1987) have recently presented some of the available multivariate techniques for 'skew-symmetry analysis' which may provide further avenues for exploring these sorts of patterns.

6.1.4   Questions remaining to be explored

These measures provide the foundation for a series of questions about community organization, which include the following.

1. How many communities are organized hierarchically?
2. Why might pairwise interactions be asymmetric?
3. Do communities vary systematically in these attributes?
4. Can we predict the traits of species higher in the competitive hierarchy?
5. How much does position in the hierarchy vary as the environment changes?

6.2   TWO EXAMPLES OF COMMUNITIES WITH COMPETITIVE HIERARCHIES

There are two sets of data where these questions can be explored. The first deals with the plant communities explored by Keddy and Shipley (1989) and the second with fly communities discribed in Gilpin *et al.* (1986).

6.2.1   Example 1. Herbaceous plant communities

Patterns in plant communities

Keddy and Shipley (1989) found highly significant hierarchical organization in published matrices from seven different plant communities including lake-shores, sea-cliffs and chalk grassland. They also measured asymmetry and found that these seven matrices had highly asymmetric interactions. There was one exception, an eighth matrix from the experiment by Harper (1965) exploring competition among six varieties of flax. This exception is discussed later. Similar evidence is available from studies of intraspecific competition in plant monocultures. Weiner and Thomas (1986) reviewed sixteen published studies and found that fourteen of them supported the hypothesis that competition between plants is asymmetric.

Asymmetric competition and plant size

Current understanding of the mechanisms of plant competition may explain these observations of asymmetric competition and hierarchical organization. One of the basic characteristics of plants is their dependence upon sunlight. If two plants are neighbours and one plant is slightly taller than the other, then there are two immediate consequences. The taller plant intercepts more light as a consequence of its height, thereby allowing further growth. Simultaneously it deprives the shorter plant of some photons, thereby inhibiting the growth of the shorter plant. This sets up two positive-feedback loops which can often lead to the increased success of the large plant and declining success for the small one. Short plants simply do not shade tall ones to the same extent that tall ones shade short ones. Competitive hierarchies may therefore be inevitable consequences of differences in size (Weaver and Clements, 1929; Weiner and Thomas, 1986; Keddy and Shipley, 1989).

Of course, plants do not compete only for light, nor do they interact only by shading. However, access to light can control acquisition of other resources. Light is necessary both for constructing roots and for the physiological processes of nutrient uptake. Plants with more access to light will therefore be better able to forage for nutrients than plants with less access to light. Plants which are shaded are not only denied the photons necessary for constructing above-ground parts, but, simultaneously, their access to raw materials is reduced, thereby further restricting growth. Minor differences in height can have a major effect on both the quantity and the quality of the light available. Fitter and Hay (1981) have shown that light availability declines exponentially with distance below the top of the canopy; also, the red to far-red ratio declines below the canopy. Weiner (1986) designed an elegant experiment to compare above- and below-ground competition; he found that competition for light was asymmetric whereas competition for nutrients was symmetric. Other relevant examples are discussed in Keddy and Shipley (1989).

Traits conferring competitive ability

If asymmetric interactions for light control the position in competitive hierarchies, then plant height should be significantly correlated with position in the competitive hierarchy. Clements (1933) summarized the results of hundreds of transplant and removal experiments in prairie vegetation (for example, Weaver and Clements, 1929; Clements et al., 1929) and concluded that, in general, 'the taller grasses enjoyed a decisive advantage over the shorter'. Goldsmith (1978) studied sea-cliff plants and showed that the larger species suppressed the smaller (Fig. 6.2). S. D. Wilson and Keddy (1986b) experimentally derived a competitive hierarchy for seven shoreline species. The dominants were tall species whereas the subordinate was a small rosette species; Keddy and Shipley (1989) showed that more than one-third of the competitive ability of these species in mixture could be predicted from knowledge of their heights ($r^2 = 0.37$). Similarly, in the chalk grassland study, Mitchley and Grubb (1986) derived a dominance hierarchy for 6 plant species and found a significant correlation between position in the hierarchy and mean turf height in monocultures; Mitchley and Grubb noted that 'the plants with the tallest leaves were the most effective in interference'. Mitchley (1988) has since shown that there is a positive correlation between the height of grassland species and their relative abundance.

Since diallele designs increase in size by the square of the number of species examined, there are upper limits upon the number of species which can be studied to relate traits to competitive ability. To overcome this problem, Gaudet and Keddy (1988) used a modified additive design incorporating 44 wetland plant species differing greatly in size, morphology and other traits. To measure competitive ability, each plant species was grown with a phytometer (Lythrum salicaria), and competitive ability was measured as the ability to suppress this phytometer. They showed that simple traits such as biomass, height and canopy diameter could account for 74% of the measured competitive ability. Above-ground biomass was the best predictor ($r^2 = 0.62$). Height was also significant ($r^2 = 0.43$). A subset of the species was tested against a different phytometer and similar results were obtained. Figure 6.3 plots the percentage reduction in the phytometer against the above-ground biomass in 44 species.

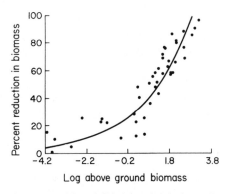

Figure 6.3 Screening for competitive ability in wetland plants. Percentage reduction in the biomass of *Lythrum salicaria* (when grown with different neighbours) plotted against the mean above-ground biomass of each of the neighbours (44 species, each point is the mean of $n = 5$ replicates, $y = \exp(3.34 + 0.44x)$, $r^2 = 0.69$). The points on the left represent small species such as *Ranunculus reptans* and *Lobelia dortmanna*, whereas the points on the right represent large leafy species such as *Typha latifolia* and *Lythrum salicaria* (data from Gaudet and Keddy, 1988).

6.2.2 Example 2. Fly communities

Gilpin *et al.* (1986) looked at these questions, using experimental fly communities. Figure 6.1 showed evidence of hierarchical organization; Gilpin *et al.* present five matrices from different environments (food types and temperatures) which demonstrate hierarchical organization visually.

Gilpin *et al.* also asked whether traits measured on individual populations could predict position in the hierarchy. Taking the largest matrix in Fig. 6.1 (thick food, 25°C), they compiled 11 independent variables which might be predictors of competitive ability. Four variables accounted for 80% of the variance in competitive rank. These variables were: (1) a measure of the exponential rate of increase; (2) a measure of carrying capacity; (3) a measure of maximum daily rate of emergence; and (4) larval production. Thus, they concluded, if a new fly species were added to the experiment, then its rank in the competitive hierarchy could be accurately predicted from these measurements made on populations explored in isolation.

Using the four largest matrices, they also pose the question of variation in hierarchies among environments (Table 6.1). For example, when temperature

Table 6.1 Competitive ranks[a] of 20 species of *Drosophila* as determined in four different environments represented by four combinations of temperature and food thickness (from Gilpin *et al.*, 1986, Table 2.4). Although individual species do shift competitive rank, there is highly significant concordance of the hierarchies across the four environments (Kendall's $W = 0.61$, $P < 0.001$). There is also a suggestion that hierarchies vary with changing food thickness more than with changing temperature

25°C, thick	19°C, thick	25°C, thin	19°C, thin
1	1	3	6
2	2	1	5
3	4	10	4
4	5	11	1
5	7	13	12
6	6	2	2
7	17	6	11
8	12	5	3
9	8	18	16
10	20	4	20
11	3	20	14
12	14	9	15
13	10	8	10
14	18	12	8
15	11	7	13
16	13	19	18
17	19	15	17
18	15	16	19
19	16	14	7
20	9	17	9

[a] 1, best competitor; 20, worst competitor.

changed they found many cases where position in the hierarchy shifted by four or more competitive ranks. In one case (*Pallid*) a species shifted from position 20 to position 9 as temperature dropped from 25 to 19°C. Thus, at the fine scale, hierarchies in fly communities are not invariant. On the other hand, on a coarser scale, there are impressive similarities among the four environments. In the foregoing example the two communities still have correlated ranks in the hierarchies (Spearman rank-correlation test, $P < 0.01$) and, over two food types and two temperatures, there is strong concordance (Table 6.1).

Based on such studies, Gilpin *et al.* conclude that assembly rules exist; there were 2^{10} ($= 1024$) different possible combinations of their 10 species, but they found only 2 to 7, depending upon which states were considered stable and transitional. However, they add, even in the highly simplified laboratory environment, and with 8 years of work, they still have not established the proximate mechanisms of competition. That is, their understanding of what produces the hierarchy is rudimentary. On the positive side they were able to demonstrate the existence of assembly rules based upon competitive hierarchies, and take the step of predicting competitive abilities from measured traits of the species.

6.3 FUTURE DIRECTIONS

The possible avenues for this sort of research are largely unexplored. One constraint may be the scarcity of published matrices, but this can be rectified if the demand exists. Another constraint may be the lack of quantitative measures of matrix structure; the study of food webs, for example, progressed rapidly once quantifiable attributes were defined and patterns in these attributes were explored (Pimm, 1982). Consider some of the questions that remain to be addressed using community matrices.

6.3.1 Invariance of hierarchies

The above examples suggest that reversals of rank order in competitive hierarchies should be relatively uncommon, except where the organisms concerned are similar in competitive ability. This would be the case if the hierarchy resulted from only minor differences in competitive ability, or if a pair of organisms with similar positions in a hierarchy interacted. It is widely assumed (although rarely explicitly stated) that pairs of species do differ only slightly in competitive ability, but this assumption may follow from the tendency of ecologists to examine pairs of similar species. A consequence is that minor changes in the experimental environment are very likely to determine which species will competitively exclude the other. If we consider mixtures of very different species, however, the outcome of competitive interactions may be much more predictable: the more asymmetric the interaction is, the more likely the outcome will be relatively independent of the environment (Fig. 6.4). Communities with very different organisms may

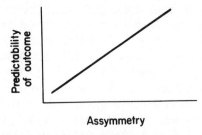

Assymmetry

Figure 6.4 The predicted relationship between asymmetry of competitive interactions and the predictability of the outcome. That is, when species are similar in competitive ability, the outcome may be most sensitive to minor changes in environmental conditions. Conversely, robust predictions are more likely to be obtained for communities that are organized in competitive hierarchies.

therefore have behaviour which can be predicted accurately in spite of environmental fluctuations. This prediction is clearly testable.

6.3.2   Traits associated with competitive ability

All of the above examples found evidence that certain traits conferred competitive ability in the populations examined, although only the plant examples had an obvious mechanism to account for these correlations. Many other studies suggest that certain traits are correlated with competitive ability. Size is an attribute which is cited for both animals (for example, Miller, 1967; Buss, 1980; Persson, 1985) and plants (for example, Grime, 1979; Gaudet and Keddy, 1988). Tilman (1982) suggests that the species which can produce the lowest resource levels will be the competitive dominant; others (Grime, 1977, 1979; Crick and Grime, 1987) propose that foraging ability is an essential component of competitive ability (see also Grace, 1989). The above examples show how we can actually test such hypotheses.

6.3.3   Attributes of communities

Do communities differ significantly in their hierarchical structure? In their mean levels of asymmetry? In their mean intensity of competitive interactions? If so, are there predictable patterns? Does asymmetry increase with resource levels? With increasing variation in plant size? If we assign species to life-forms (du Reitz, 1931), guilds (Root, 1967), life-history types (van der Valk, 1981), strategies (Grime, 1979) or functional groups (Day et al., 1988), do we find that within-group interactions are symmetric and among-group interactions are asymmetric?

6.3.4   Asymmetry, coexistence and the competitive exclusion principle

Since Hutchinson's (1959) paper on coexistence, entitled 'Homage to Santa Rosalia', ecologists have expended large amounts of energy addressing the 'problem' posed by the coexistence of similar species. However, if we postulate that communities often have inclusive niches and competitive hierarchies, then it is the coexistence of very different species which becomes the problem! Similar species could coexist precisely because interspecific competition is approximately equal to intraspecific competition, thereby weakening inter-specific interactions which might otherwise lead to exclusion (Aarssen, 1983, 1985). Minor environmental fluctuations might allow nearly equivalent species to persist indefinitely. The real dilemma then becomes how to explain the survival of mixtures of species, particularly subordinate species in mixtures containing competitive dominants. This deserves more-detailed consideration, since these ideas relate directly to the ever-popular competitive exclusion principle; the remainder of this subsection explores the competitive exclusion tangent in more detail.

One could ask a variety of questions about competition and community assembly, but one question which has received overriding attention is the role of competition in controlling the number of species which can coexist. This interest can be traced back at least to Hutchinson's (1959) paper, but probably extends back at least to Victorian times. In fact, a central theme of Darwin's *On the Origin of Species by Means of Natural Selection* (1859) is how biological diversity arose. It is a simple extension of this idea to ask how the biological diversity of a particular site is maintained. Moreover, it fits nicely with that part of the human psyche which likes collections, whether they be of birds' eggs, butterflies, stamps or Elvis Presley memorabilia, and a first question for any collector is always 'How many do you have?'. We might argue that the question of coexistence is not the most pressing in community ecology, but let us for the moment accept the historical fact and look at approaches which have been used to address the question.

The conventional view starts with the assumption that niche differentiation is the primary principle around which communities are organized. If this is correct, then the coexistence of similar species becomes difficult to explain. Species with similar niches should be least likely to coexist through time, because of the intense competition between them. Darwin (1859) wrote:

> ... it is the most closely allied forms – varieties of the same species and species of the same and related genera – which, from having nearly the same structure, constitution and habits, generally come into the severest competition with each other.

Hardin (1960) expanded on this idea and called it the **competitive exclusion principle**.

The basic approach of experimental studies of this principle has been to choose two closely related species, culture them together, and ask whether they are capable of coexisting. If they do not coexist, then it is clear that they were 'too similar' for niche differentiation to allow coexistence, and if they do coexist, then one may be able to find the axis along which they coexist. Following Darwin's observation, it is frequently assumed that closely related species occupy similar niches, and that therefore pairs of coexisting closely related species are of particular interest (for example, Harper, 1960). Harper and his co-workers, for example, did a long series of studies entitled 'The comparative biology of closely related species living in the same area' where closely related species were allowed to compete (for example, Harper and Chancellor, 1959; Clatworthy and Harper, 1962; Harper and McNaughton, 1962; Harper and Clatworthy, 1963). Moreover, the de Wit replacement series, one of the basic techniques for exploring competitive interactions in mixtures (Harper, 1977), absolutely requires that the two species be similar in size in order to avoid confounding species composition with changes in 'density' (more strictly, biomass). The very fact that papers using this design refer to maintaining constant 'density' rather than 'biomass' reveals the fundamental requirement that the species should be similar in size.

The general conclusions of most such studies follow those of Gause's classic experiments with yeast (1932) and *Paramecium* (1934) populations and of Park's experiments with *Tribolium* beetles (Park, 1948, 1954): pairs of experimental populations rarely coexist indefinitely. This has led to the general acceptance of the competitive exclusion principle, which Hardin (1960) describes and summarizes as 'ecological differentiation is the necessary condition for coexistence'.

Although this is widely accepted as an axiom of community ecology, it actually tells us very little. Unless there is some way of specifying how similar species must be to coexist, then there is no way of refuting the principle. Although the point has been made before (Miller, 1967), it is worth repeating. If the species exist, then they are sufficiently different; if they do not coexist, then clearly they are too similar. The argument is thus completely circular. It has the pleasing result of being able to explain whatever results we find, but little ability to predict them. An experiment which combined groups of similar species and groups of dissimilar species and followed coexistence through time might be able to falsify the principle; on the other hand, the results could always be criticized because an inappropriate measure of similarity may have been chosen. Devising an objective measure of similarity might be one way to bring the principle back into the realm of science. However, Hardin (1960) argues that the principle is no more testable than are Newton's laws, and that by demanding that the principle be testable we miss the point. Perhaps he is right, but this leaves the unsettling question: what, then, is the point of growing pairs of related species in experiments and invoking the principle to explain whichever outcome is found?

The competitive hierarchies illustrated in this chapter challenge this assumption that competition is most intense among similar species (for example, MacArthur, 1972; Cody, 1974; Diamond, 1975; Werner, 1979; Pianka, 1981). It may be true that competition is most *symmetric* among closely related taxa, but it is arguable that similar and symmetric competitive abilities should lead to long-term coexistence, since neither species will have a clear advantage over the other. In fact, this very point has been made elsewhere (Aarssen, 1983; Agren and Fagerstrom, 1984; Goldberg, 1987). In contrast, among very different taxa, competition may be highly asymmetric and intense. The logical conclusion is that species which are different may be unlikely to coexist, since their interactions are likely to be asymmetric, and exclusion should occur more rapidly than when interactions are symmetric! This leads to a prediction that is exactly opposite to the competitive exclusion principle: as species become more different, so they are less likely to coexist. The competitive exclusion principle is therefore not only circular and untested, but there are perfectly good reasons for a contrary prediction.

Figure 6.5 sketches a possible reconciliation of these contrasting predictions regarding competition and coexistence. At one end (left) there is the limiting

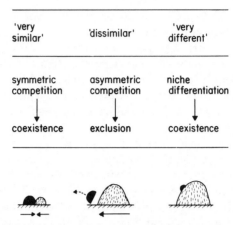

'very similar'	'dissimilar'	'very different'
symmetric competition	asymmetric competition	niche differentiation
↓	↓	↓
coexistence	exclusion	coexistence

Figure 6.5 Does similarity produce competitive exclusion or coexistence? The text argues that there are reasonable arguments for both predictions, even though the former is widely accepted and is called 'the competitive exclusion principle'. This figure suggests that the answer in part depends upon just how similar two species are. If the species are very similar (left) there is coexistence because the species are so similar that competitive abilities are equivalent and symmetric. Decreasing similarity leads to dominance and competitive exclusion (middle). If the species are sufficiently different, then there is coexistence through resource partitioning or amensalism. The bottom figures represent the situations pictorially by sketching possible interactions between barnacles of different sizes on a rock surface; the solid arrows indicate the effect of each competitor upon the other.

case where very similar species coexist precisely because they are so similar. As the organisms become increasingly different, so the possibility of asymmetric competition increases, leading to dominance and competitive exclusion (middle). As the organisms become sufficiently different they coexist by means of niche differentiation (right). Recalling Connell's work discussed in Chapter 5, we can imagine an assemblage of sessile marine invertebrates such as barnacles; the first case (left) shows interactions between two species which are morphologically similar, the second case (middle) shows the effect of a large species upon a smaller species, and the third case (right) shows the relationship between a large species and a small, attached amensal species. This reconciliation would require objective and operational measures of similarity (for example, Green, 1980; Legendre and Legendre, 1983). Here is where the one exception in the review by Keddy and Shipley (1989) becomes relevant. Recall that there was one matrix where there was little evidence of either asymmetry or competitive hierarchies. This matrix, from Harper (1965) was based upon competitive interactions among varieties of a single species. This could well be an example of the situation in Fig. 6.5(left). Recent work on competitive hierarchies in four pasture plants (Aarssen, 1988) may also illustrate this situation.

An alternative approach to a synthesis might require explicit consideration of resources. Perhaps one outcome occurs in space-limited communities of sessile organisms, and another in niche-differentiated communities where populations are limited by resources other than space. This distinction is discussed further in Chapter 7.

6.4 CONCLUSION

Studies of competition have emphasized pairwise interactions between small numbers of populations and the way in which these outcomes vary with changes in environment (for example, Gause 1932, 1934; Park, 1948, 1954). These have become classic studies cited in undergraduate textbooks (for example, Krebs, 1978; McNaughton and Wolf, 1979; Pianka, 1983). In contrast, the studies explored here have considered interactions among many pairs of species simultaneously. Rather than strong environmental dependence, general principles about hierarchical organization are suggested. It is too early to reconcile these apparent differences. However, it is clear that scientists will not find what they do not seek. Exploring the environmental dependence of interactions between one pair of very similar species (or genotypes!) does not provide the possibility of discovering hierarchical organization or general relationships between competitive ability and species traits. Perhaps when experiments are designed to look for complexity, complexity is found, but when they are designed to look for general principles, general principles are found, too. These different kinds of laboratory experiments certainly present very different views of what is possible in nature.

The above studies of Gilpin *et al.* (1986) and Keddy and Shipley (1989) illustrate only a small fraction of the possibilities which exist for studying community matrices. Although many of the existing examples come from the artificial conditions of laboratory experiments, there are examples from the field (Buss, 1979, 1980, for marine sponges; Siefert and Siefert, 1976, for insect communities in *Heliconia* flowers; Mitchley and Grubb, 1986, and Fowler, 1981, for plants). The exploration of the properties of such matrices, and the construction of larger matrices representing different environments and systems, is one way to answer some of the unresolved questions posed in this chapter. Some of these questions are the following. (1) How many and what kind of communities are organized by competitive hierarchies? (2) Can we predict to what degree an interaction will be asymmetric? (3) Are there measurable traits which allow us to predict the position of species in a competitive hierarchy? (4) How much do hierarchies and the traits associated with them vary among environments? There is much to be done.

6.5 QUESTIONS FOR DISCUSSION

1. How do the results of classical studies like those of Gause and Park differ from those presented here? Why?

2. Consider the methods available for constructing matrices of competitive ability. What are the strengths and weaknesses of each?

3. Can you find other published examples of experimentally derived matrices? Are they consistent with the patterns presented here?

4. Are there other questions which can be asked about such matrices?

5. Consider the cost of constructing large matrices. Does the screening technique used by Gaudet and Keddy (1988) offer an alternative route for exploring such questions more cost-effectively?

6. Why have ecologists placed so much emphasis upon comparing pairs of similar species? What, if anything, have we learned from such studies?

7. Consider the model sketched in Fig. 6.5. Are there better ways of dealing with the contrary predictions about similarity and coexistence?

7 Competition, empiricism and comparison

... the better competitor may exclude the other
species even though in a habitat where both normally
coexist an observer might only witness severe
competition 1 year in 20. This is the reason most
evidence for competition is from biogeographers.

R. H. MacArthur (1972)

... ecologists are unusual among scientists in that
they deny the need for empirical science.

F. H. Rigler (1982)

The role of competition in nature may be best explored and summarized using
general models which compare across species and across habitats. This
requires three steps. First, one must choose measurable dependent variables
such as the intensity or asymmetry of competition which describe one or more
attributes (or states) of a system. Secondly, one measures independent (or
predictor) variables that one anticipates may be useful in predicting the
dependent variables. Thirdly, one explores relationships among these variables. It should be emphasized here that the objective is not to produce vague
ideas that environments which differ in unmeasurable ways differ in certain
unmeasurable properties, although such speculations often pass for theory.
There are simple quantitative techniques available to find and test for robust
correlations among state variables in ecological systems (R. H. Peters, 1980a;
Rigler, 1982). Lewontin (1974) attributes current problems in community
ecology to the lack of agreement about important state variables, and Rigler
(1982) suggests that lack of consensus may explain why empiricism has not
been widely accepted. The general empirical relationship between mean size of
individuals and density described from plant monocultures (the $-3/2$ law:
Harper, 1977; Westoby, 1984) illustrates two state variables and an empirical
relationship which has been of broad general interest to population biologists.
Keddy (1987) observed that once plant community ecologists agreed upon two
state variables and explored diversity–biomass relationships in different

vegetation types, important generalizations were also discovered. There are growing numbers of examples of this approach (for example, Gorham, 1979; Damuth, 1981; Brown and Mauer, 1986; Currie and Paquin, 1987).

It is not yet clear which dependent and independent variables are important for the development of competition theory. For studies of intraspecific competition, performance and density are obviously of greatest interest. For most other questions and levels of organization, choosing state variables and exploring their interdependence will often rest upon the comparative approach (Chapter 4). Several paths exploring these relationships already exist, and others may be found. For example, if it were possible to quantify the abundance of resources, or their distribution in time and space (Chapter 1), then one could relate these to the intensity of competition in specific communities (Southwood, 1977, 1988; Price, 1984a; Schoener, 1986). One could also compare different kinds of organisms to identify traits which may confer competitive ability (Grime, 1977). This could be made more rigorous by systematically screening for traits of organisms (Rorison et al., 1987) or even by developing bioassays of competitive ability (Gaudet and Keddy, 1988). One might also use field experiments to measure the effects of competition in different habitats and try to correlate these effects with different environmental variables (Connell 1961; del Moral 1983; S. D. Wilson and Keddy, 1986a; Gurevitch, 1986).

When using such empirical comparative approaches, it is important to remember that competition may be very important in controlling the evolution, distribution and abundance of species even if it occurs only occasionally (MacArthur, 1972, quoted above; Weins, 1977). The failure to observe competition in a specific study says only that it did not occur under the precise conditions of that study, and not that it never occurs in that habitat or with that species. Empirical comparisons across communities and species, exploring the relationships between the traits of organisms and their environments, may be one of the few ways of surmounting this difficulty.

By determining how competition intensity varies among environments ecologists could also place boundary constraints on other kinds of competition models. The issue is not whether competition, resource patchiness or predation alone controls species abundances and community competition. Rather, the problem is to define regions where these forces operate. These regions must be operationally defined; i.e. measurable. To reach this degree of sophistication in competition models it will be helpful to compare different systems and different species. Table 7.1 lists some recent studies which have used this approach.

The examples which follow illustrate the search for general principles. The challenge is always to devise predictive relationships which can be tested, as opposed to *post hoc* explanations for patterns which have already been found. After surveying trends independently in sessile organisms (plants) and motile

Table 7.1 Examples of studies exploring effects of different kinds of environments upon competition and related aspects of population and community ecology

Source	Environmental variables	Dependent variables
Grime (1974, 1979)	Stress	Competition
	Disturbance	Plant strategies
Connell (1975)	Predation	Diversity
	Harshness	Competition
Huston (1979)	Rate of displacement	Diversity
	Frequency of reduction	Competition
Price (1984a)	Types of resources	Importance of
	Rates of population	competition
	responses	Operation of alternative
	Spatial distributions	organizing forces
Southwood (1977)	Favourableness of	Life-history traits
	environment	
	Temporal variation	
	Spatial variation	
	Generation time	
	Foraging range	
Southwood (1988)	Disturbance	Physiology
	Adversity	Defence against
	Biotic interactions	predators
		Food harvesting
		Reproductive activities
		Escape tactics

organisms (insects), possible convergent trends are discussed. The chapter concludes by returning to the topic of research strategies.

7.1 CONSTRAINTS ON COMPETITION IN PLANT COMMUNITIES

What controls the importance of competition in organizing plant communities? This is a topic in plant ecology which has produced acrimonious debate and not a little rhetoric. Let us return to the question of coexistence and the Lotka–Volterra model of Chapter 3 for a starting point. Figure 7.1(left) shows the conditions for competitive exclusion. What could prevent competitive exclusion in this simple model system? The first, and most obvious answer is something which continually reduces population sizes to prevent equilibrium from being obtained. This model only predicts extinction if equilibrium occurs; the populations coexist until the trajectory collides with an axis, and this can be prevented by repeated harvesting (Fig. 7.1, centre). A second

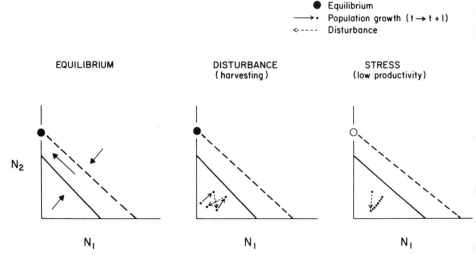

Figure 7.1 Constraints upon competitive exclusion illustrated in a simple two-species Lotka–Volterra system where there is competitive dominance and therefore exclusion at equilibrium (left). If there is repeated disturbance (middle) or occasional disturbance with very low rates of recovery (right), then two species can coexist indefinitely. There are therefore many possibilities for coexistence which are not 'equilibrium coexistence' in the sense of the Lotka–Volterra equations (Fig. 3.2), but the word coexistence is often carelessly used to describe the much narrower concept of 'equilibrium coexistence'.

possibility is to consider the rate at which population sizes change. If population growth rates are extremely slow, then clearly, after a single harvest, the populations can coexist for long periods as they gradually approach an axis; the slower the rates of population growth are, the longer the two populations can coexist (Fig. 7.1, right). Periodic harvests, combined with relatively slow rates of growth, can generate non-equilibrium coexistence for virtually unlimited periods. These are but two ways in which the competitive exclusion predicted by the Lotka–Volterra model can be avoided; Huston (1979), Chesson and Case (1986) and Chesson (1986) have explored these and other non-equilibrium explanations for coexistence. Equilibrium coexistence may therefore be the least interesting and least likely mechanism of coexistence, even if it is the most mathematically elegant (Chapter 3).

What shall we call the above two forces acting to reduce rates of competitive exclusion? Huston (1979) used the Lotka–Volterra model to explore non-equilibrium interactions of this sort. He produced a model exploring the interaction of these two forces, and called them, logically enough, (1) the rate of competitive displacement, and (2) the frequency of reduction.

Grime (1977, 1979) reached similar conclusions, using very different evidence. He was concerned not with specific environments, nor with coexistence, but rather with the evolutionary responses of plants to their environments. He asked what basic kinds of plant life-history types can be found, and proposed three major sets of correlated traits, or 'strategies'. According to Grime, one such set of traits evolves under relatively benign conditions where plants are selected for the ability to compete for contested resources. Foraging ability is the key trait under such conditions. He then postulates two environmental factors which have constrained the development of such traits. One is disturbance, which Grime defines simply as the removal of biomass from plants. Southwood (1987, 1988) calls this durational stability, the length of time for which the habitat is available scaled to the generation time of the organism. The other is stress, which Grime defines as environmental factors which reduce the rate of growth of plants – essentially the inverse of primary productivity. This has been called adversity (Greenslade, 1983; Southwood, 1987, 1988) and described as conditions 'that threaten the homeostasis of cytoplasm, that interfere with the normal functioning of the enzyme systems, or that destroy the integrity of membranes' (Southwood, 1987). Although devised for different purposes and assembled from different data, there is a remarkable convergence with the Huston model.

Both of the above models suggest that communities can be arranged in a two-dimensional array. One axis is labelled disturbance, and along it communities are arranged from those with minimal disturbance to those with frequent and intense disturbance. This axis may need to bifurcate to accommodate both different frequencies and intensities of disturbance (Miller, 1982; Sousa, 1984). The other axis consists of communities with increasingly slow rates of recovery from disturbance; i.e. increasingly low productivity. The distribution of communities along this axis raises an apparent paradox. Let us first consider the consequences of disturbance, and then return to the paradoxical consequences of stress.

7.1.1 The effects of disturbance on competition

Although disturbance has received a great deal of recent attention in ecological studies (for example, White, 1979; Sousa, 1984; Pickett and White, 1985; Huston, 1985; Mann, 1985), few field experiments have been undertaken to test whether and measure how the effects of competition decrease as disturbance increases. Such experiments would need to compare areas differing in the amount of disturbance (either natural or experimentally produced) and measure the number and intensity of competitive interactions. One example of such a study was attempted by Dayton (1975), where a dominant species of macro-alga was removed from plots in the rocky intertidal zone on the west coast of North America. By comparing sites with different amounts of exposure to waves, Dayton tried to obtain sites with different amounts of natural disturbance. Figure 7.2 compares two sites, one with and the other without wave damage. In both cases the removal of the canopy

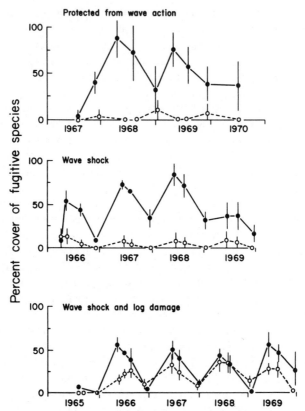

Figure 7.2   Effects of disturbance on the intensity of competition in a rocky intertidal plant community (Dayton, 1975). The open dots are control plots and the solid dots are plots where the canopy of *Hedophyllum sessile* was experimentally removed. The difference in percentage cover of fugitive species between cleared and uncleared plots can be considered to be a bioassay of the intensity of competition experienced by the fugitive species. Note that this difference declines from protected sites (top) to sites exposed to waves and log damage (bottom).

species, *Hedophyllum sessile*, resulted in release of a group of fugitive species of algae. There was no obvious difference between the two wave exposures (Fig. 7.2, top and middle). The low levels of abundance of fugitive species in the control plots suggest that the site exposed to wave shock really was not disturbed much more than the first site. Dayton provides only a verbal assessment of the different levels of exposure, and it is not clear whether the macro-algae actually detect the differences which appear obvious to the human observer. The effects of *Hedophyllum* removal are much less dramatic in the area exposed to log damage (Fig. 7.2, bottom), and therefore, presumably, the intensity of competition between *Hedophyllum* and the fugitive species is lower. Fugitive species were already more abundant at this site, suggesting that the algal community was indeed responding to the presence of a natural disturbance.

Gradients of competition can also be found by comparing patches in vegetation. For example, Platt and Weis (1985) explored plant competition in prairies among fugitive plants which occupy open sites produced by badgers. They showed that clearings which were already colonized by one species were less suitable for colonization by a second species (Table 7.2). Therefore, open sites could be ranked in order of suitability to colonists. In the case of *Mirabilis hirsuta*, for example, unoccupied open sites were most suitable, with sites occupied by *Solidago rigida*, *Verbena stricta* and *Mirabilis hirsuta* being increasingly unsuitable for colonists.

Table 7.2   The responses of three prairie plant species to sharing disturbed sites with a second species. The dependent variables measuring performance were seed production and plant height, both of which are important traits for dispersal to new disturbed sites (after Platt and Weis, 1985, Table 2). For each species there was a gradient of effects of competition depending upon the co-occupant of the disturbed sites

Species	Neighbour	Seed production (no. plant^{-1} yr^{-1})	Height (cm)
Mirabilis hirsuta	–	150	75
	Solidago	140	71
	Verbena	130	67
	Mirabilis	120	62
Verbena stricta	–	330	65
	Solidago	220	60
	Mirabilis	200	56
	Verbena	180	50
Solidago rigida	–	460	57
	Verbena	270	48
	Mirabilis	205	43
	Solidago	185	40

7.1.2   The paradox of resource limitation

In communities with low productivity there will be relatively slow rates of recovery from perturbation. Consequently, large numbers of species may be able to coexist (Fig. 7.1, right) because equilibrium and exclusion do not occur before another (infrequent) disturbance occurs. We can develop one line of argument (for example, Huston, 1979) that leads to the prediction that the effects of interspecific competition should increase with resource levels. Imagine a field experiment measuring competition intensity in communities with different amounts of stress. One might predict that as stress decreased, rates of recovery would increase, and this would in turn increase the opportunities for individuals both to encounter and to interact with neighbours. Therefore competition intensity would increase with decreasing stress. If disturbance also increased with stress, these effects would be magnified since the vegetation with the lowest rate of recovery from disturbance would be the one most frequently disturbed. This is the case, for example, on lakeshores, where increasing disturbance from waves produces declining soil fertility. S. D. Wilson and Keddy (1986a) found that competition intensity increased with increasing fertility and decreasing disturbance in their field experiment measuring competition intensity along a natural environmental gradient (Chapter 2).

However, here is the paradox. Chapter 1 proposed that resource limitation was essential for competition to occur. Competition should therefore be most intense in the sites where resource levels are lowest. That is, the very definition of competition leads us to expect exactly the opposite of the line of reasoning developed above, since otherwise we are left with the observation that competition increases as resource levels increase! Areas with obvious resource limitation, such as arid and semi-arid areas, would appear to be ideal for testing such ideas. Fowler (1986) observes that although plant competition in such habitats is more often assumed than demonstrated, there is some evidence that competition may be higher in areas with lower water status.

The resolution of this paradox is one of the more challenging questions in plant community ecology (Grime, 1973; Newman, 1973; Thompson, 1987; Tilman, 1987b; Thompson and Grime, 1988; Grace, 1989). It is unlikely that the issue will be resolved by debate; we need critical tests for differentiating between these contrasting predictions. There is no reason why some clear, contradictory predictions cannot be made, and the conclusive experiments run.

For such experiments to be defined and run, there has to be agreement upon what the real issues of debate are. As discussed in Chapter 1, different resources may produce different kinds of competitive interactions and community structure. One critical experiment (Weiner, 1986) compared competition for light with competition for nutrients and found that competition for light was asymmetric, whereas competition for nutrients was not. J. B. Wilson (1988) has reviewed the literature on root and shoot competition in agricultural species. He concluded that root competition is usually more important than shoot competition, and that adding resources may either increase or decrease competition. Since most agricultural species are short lived and occupy relatively fertile and disturbed sites, the implications of these results for other plant communities are not clear.

Resource consumption is not the only important factor to consider in plant competition, since growth rates and survival of populations will also be determined by the efficiency with which they use and conserve nutrients.

Chapin (1980) reviewed the nutrition of wild plants and observed that 'adaptations that minimize nutrient loss ... have received less attention than adaptations related to nutrient absorption and growth'. Similarly, models of competition usually emphasize rates of population growth and rates of resource consumption (Chapter 3) rather than conservation. Resource loss rates from individuals get hidden in the r-term (Lotka–Volterra model) or are assumed to be constant (Tilman, 1982). In fact, loss rates have a number of properties which can influence studies which compare competition in different habitats. One of the most important of these is the observation by Chapin (1980) that adaptations to infertile sites usually do not involve increased efficiency of resource extraction (i.e. enhanced root absorption capacity) so much as increased conservation (i.e. evergreen foliage). Incorporating resource conservation into competition studies will require explicit consideration of resources loss rates. A complication is that loss rates may be summarized for entire communities, for populations or for individuals – with different consequences for competition. Discussions (and measurements) of 'loss rates' can be confusing if one person considers the proportion of standing crop removed averaged over all species (e.g. effects of fire or grazing) and another measures differential loss among established adults (e.g. leaching from foliage). For example, Tilman (1988) has modelled the changes in plant morphology which would be expected along a loss-rate gradient where the losses are a density-independent mortality that fall equally upon all individuals independently of their size and allocation pattern. The model predicts that, if soil nutrient supply rates are sufficiently high, then higher loss rates will favour species that allocate a higher proportion of their biomass to leaves and a lower proportion to stems, which is consistent with what we know about the traits of ruderal species (Grime, 1977). We do not know what the consequences would be if variation in loss rates among species were incorporated in the model. Berendse *et al.* (1987a, b) studied nutrient conservation in two species typical of infertile pastures (*Molinia caerulea* and *Erica tetralix*), and concluded that the retranslocation of nutrients (conservation) was the principal factor explaining the competitive dominance of *M. caerulea*.

Since loss rates may also vary among habitats independently of uptake rates, measuring loss rates of different species along fertility gradients will require more attention. This is complicated by the problem that the *costs* of identical loss rates may vary among environments. In general, the lower resource levels are, the more difficult it may be for a plant to replace lost tissue, so a 'cost of loss' gradient can be imagined. This may actually run in exactly the reverse of the actual loss gradient, since relatively small losses in infertile habitats may have the highest cost of replacement. At the same time, nutrient conservation may have other costs such as increased expenditures in the protection of foliage (Chapin, 1980; Coley *et al.*, 1985; Southwood *et al.*, 1986). Explicit consideration of nutrition and nutrient conservation (Chapin, 1980; Chapin *et al.*, 1986) may be essential to evaluate the validity of current assumptions regarding plant competition and the evolution of plant traits in different environments.

There are still unresolved issues in nutrient exploitation itself. Conflicting views of foraging behaviour need to be resolved. Grime (1977, 1979), for example, implies that ability to forage for above- and below-ground resources should be positively correlated. This can be supported by noting the physiological integration of plants, and arguing that resources acquired above ground can be rapidly translocated to assist in foraging below ground, and vice versa. Moreover, if plasticity is essential for detecting and depleting patches of resources, then there is every reason to expect that such an attribute will be correlated in both above- and below-ground parts. In contrast, the view that there are trade-offs in resource acquisition (Tilman, 1982, 1988) is also inherently reasonable. Resources allocated to roots cannot be allocated to shoots, and vice versa. Thus, there should be a fundamental trade-off between above- and below-ground foraging ability. Experiments could surely test between these two contrasting views of plant foraging (for example, Crick and Grime, 1987), but the difficulty in interpreting such experiments is illustrated by the lack of consensus on the interpretation of existing data (Thompson, 1987; Tilman, 1987a; Thompson and Grime, 1988).

7.2 CONSTRAINTS ON COMPETITION IN ANIMAL COMMUNITIES

Since insects have a vast array of life-history types and resources, they are an obvious group to explore for trends in the intensity of interspecific competition in animal communities. Arguing from first principles (see Chapter 1), the effects of competition would be expected to vary with at least three factors: the quality of the resource, its patchiness and the host-specialization of the species concerned.

If resources are of high quality, then consumers using them should have more surplus resources to expend in maintaining and defending access to the resources. At the same time, the higher the quality of the resource is, the greater the return to be gained from monopolizing access to it. This leads to the prediction that higher-quality resources will provide both the evolutionary means and the incentive for interference competition.

Testing this prediction requires a measure of resource quality. One possible measure is in energy units. Assume that energy content in kilojoules per gram is a suitable measure of resource quality. Morowitz (1968) showed that the energy content of herbaceous plants is of the order of 4 kcal (ash-free g)$^{-1}$(17 kJ g^{-1}), whereas animal materials have values closer to 6 kcal (ash-free g)$^{-1}$(25 kJ g^{-1}). Another possible measure of quality of prey is the similarity of composition of predator and prey; Southwood (1985) writes of the evolutionary hurdle produced by the 'biochemical differences between plant composition and animal requirements'. If high-quality resources promote interspecific competition, both definitions of quality produce the same prediction: predators will compete more than herbivores.

Now consider the patchiness of a resource. If the resource is spread across the landscape evenly, then one might expect relatively less competition than if it is localized in patches. If the resource is patchy, then individuals may aggregate around high-quality patches, creating strong selection for those individuals capable of monopolizing such patches. Even in the absence of interference competition, aggregation may still lower resource levels in the preferred patch, possibly producing exploitation competition. If the resource is not patchy, then there may be less to be gained by interference competition, and simultaneously less opportunity for neighbours to encounter and directly influence one another.

Testing this prediction requires a measure of resource localization. Patchiness is measurable whether we measure the abundance of a food species per square unit of land surface, or whether we convert it to energy units such as kilojoules per square metre available for consumption. In either case, if we compare two kinds of prey, plant versus animal, we may expect animals, in general, to be more patchily distributed. Compare, say, the distribution of caterpillars in a tree with the distribution of the foliage of that tree. Of course, such generalizations are coarse: granivorous or nectivorous insects might find their resource distributed more like caterpillars than foliage. Nevertheless, this suggests that predators may encounter more-intense interspecific competition than herbivores.

In general, prey quantity and quality are correlated: high-quality prey (i.e. animals and seeds) are quite patchy, whereas lower-quality prey (i.e. foliage and wood) are more widespread. Both (correlated) axes yield the same prediction: more competition among meat-eating predators than among herbivores.

Host specialization is that last factor to consider. If there is biochemical co-evolution between predator and prey so that the predator can consume only one prey species, and if each prey species supports only one species of predator, then there is clearly no need to look for interspecific competition. If the predators become less specialized, then opportunities for interspecific competition would increase. This suggests that, in general, the opportunities for interspecific competition increase either with increasing polyphagy or increasing number of monophagous species per host.

When making such sweeping generalizations about a major component of the Earth's biota, there is no point in arguing from special cases. Undoubtedly special cases can be found which are consistent with or contradictory to almost any remotely reasonable pattern. This is one of the problems with using natural history data to try to falsify or validate models. Convincing evidence must come from studies which have summarized the data for many species. Consider two extremes: phytophagous insects and parasitoids.

7.2.1 Phytophagous insects

Strong *et al.* (1984) have considered the evidence for the existence of interspecific competition in phytophagous insects. They estimate there are more than 350 000 species of them, which complicates attempts at systematic study or review. (Contrast this with 8500 species of birds or 4500 species of mammals!) The major orders of phytophages include: Coleoptera (beetles), Collembola (springtails), Diptera (flies), Hemiptera (sucking bugs), Hymenoptera (wasps and sawflies), Lepidoptera (butterflies and moths), Orthoptera (grasshoppers), Phasmida (stick and leaf insects) and Thysanoptera (thrips). Strong *et al.* restrict their review to species feeding on living tissues of higher plants, excluding groups such as wood-borers, nectar feeders and decomposers. Phytophagous insects exploit plants in four ways: external feeding, sucking from the plants' vascular system, mining and gall formation.

They consider three principal sources of evidence when assessing the importance of interspecific competition in structuring phytophagous insect communities.

1. Absence of Intraspecific Competition. They argue that interspecific competition will be unimportant unless intraspecific competition occurs, and present evidence that intraspecific competition is relatively unimportant. It occurred in sex of 31 examples that they reviewed.
2. Absence of Resource Depletion. They use 'the earth is green' argument raised by Hairston *et al.* (1960) (see also Chapter 3). Briefly, Hairston *et al.* argued that obvious depletion of green plants by herbivores is infrequent, as evidenced by the abundance of green foliage which can normally be seen. Herbivores are therefore not food-limited and therefore are not likely to compete for limiting resources. Their critics (Murdoch, 1966; Ehrlich and Birch, 1967) observed that green plants are neither universally edible nor universally nutritious, and thus the presence of plants does not demonstrate the presence of food, nor the absence of resource limitation. Slobodkin *et al.* (1967) replied that within every native environment one can find herbivores that are capable of causing extensive damage to plants. That they are present, but scarce, shows that food is largely unutilized. Thus, interspecific competition for limited resources is not a normal occurrence.
3. Case Studies. Some studies have selected systems where interspecific competition seemed likely, but it could not be detected. For example, Strong (1982a, b) studied tropical beetles on *Heliconia* plants. *A priori*, there was every reason to expect competition. The communities have many species, the species are closely related, and they eat virtually the same parts of the host plant at the same time. Yet there was little evidence for either interference or exploitation competition. Neither adults nor larvae showed hostile behaviour, even when close to each other. There was no evidence of interspecific association or segregation, nor was there evidence that such patterns changed with beetle density. The beetles lived in rolled leaves, and

they lived at densities far below the level which depletes this resource.

Thus, Strong *et al.* (1984) conclude that although competition does un-doubtedly occur in certain cases, it is not the prominent force structuring communities of phytophagous insects.

There are weaknesses in these arguments. With respect to the case study by Strong (1982a, b) it is clear that one cannot convincingly infer competition (or lack of it) by examining pattern (Chapter 4). Moreover, Strong *et al.* (1984) argue that asymmetric competition is frequently observed in insects, in which case the subordinate clearly is affected by competition. Thus, the case is not yet closed, but their general conclusion may still be correct.

7.2.2 Insect parasitoids

Insect parasitoids must offer one of the most dramatic cases of potential damage from intra- and interspecific competition. The host larvae which are parasitized represent a fixed supply of food and, with very rare exceptions, the parasitic larvae cannot move from one host to another. If too many parasitoids are present in a host, then some must die.

Salt (1961) reports that parasitoids are found in four different orders of insects, but the life-cycles share certain features. In general, the adult parasitoid seeks a host for its progeny and, having found one, lays an egg inside it. The egg hatches into a larvae which gradually consumes its still-living host from within. Vital organs, such as the digestive system, apparently remain entire and functional until the end.

This host–parasite relationship has been extensively studied, probably in part because of the disturbing theological questions for those who believe in a benevolent and omniscient Creator. From the point of view of the host, these two traits of the Creator would seem incompatible. Gould (1983) has elaborated upon this in an essay entitled 'Nonmoral nature'. Gould quotes a letter which Darwin wrote to Asa Gray in 1860:

> I own that I cannot see as plainly as others do, evidence of design and beneficience on all sides of us. There seems to me too much misery in the world. I cannot persuade myself that a beneficient and omnipotent God would have designedly created the Ichneumonidae with the express intention of their feeding within the living bodies of Caterpillars, or that a cat should play with mice.

Given the obvious resource limitations upon larvae sharing a host, one would predict that female parasitoids should be selected for the ability to detect the differences between healthy and parasitized hosts. Some parasitoids can discriminate (Vinson, 1976) and thereby refrain from raising the popul-ation density within the host. The ghost of competition past (Connell, 1980) can therefore be raised as a possible cause of behavioural traits of female parasitoids.

The ability to detect parasitized larvae presents an evolutionary problem which would seem to be solvable: a variety of behavioural or chemical cues in the host might serve as an indicator of the presence of parasitoids. The situation is complicated in the case of gregarious parasitoids, however. If they are to avoid competition yet efficiently exploit their hosts, then the female must not only recognize parasitized hosts, but estimate both the number of parasite progeny already present and the capacity of the host. Salt (1961) provides some limited evidence of such behaviour. It is now known that parasitoids leave chemical markers after oviposition, and these markers inhibit further attack (Vinson, 1976).

Of course, such *post hoc* explanations for behaviour are weak evidence for past competition, and experiments for measuring competition intensity in the past are difficult to design. However, competitive interactions in parasitoids can be observed directly, since when the female fails to make a perfect series of decisions, additional progeny may find themselves sharing the same host. In this case there are some interesting parallels with sessile organisms such as vascular plants or corals, since neighbouring individuals cannot avoid competition by moving. In the case of solitary parasitoids there are two principal means of interference competition (Salt, 1961; Askew, 1971): physical attack and physiological suppression. The former, writes Salt, 'has often been called fighting, but... in some cases is so one-sided as to have more in common with thuggery'. The less-sanguinary, but apparently equally efficient, alternative is for the dominant parasitoid to secrete a substance specifically responsible for the suppression of competitors. These are excellent examples of the mechanisms of dominance discussed in Chapter 1.

More-complex mechanisms may also produce competitive dominance. Fisher (1961) studied interspecific competition in parasitoids by creating two-species mixtures within prey. He inserted a single egg of each of two species of ichneumonids into caterpillars, using a micropipette. The larvae usually met within 1 day of hatching and fought, one being eliminated. The mechanism was interesting – the loser was usually not killed outright, but simply injured and thereby rendered susceptible to encapsulation by the host. If the eggs were inserted at different times, then the older of the two larvae was always the winner.

7.3 COMPARISON OF INSECT AND PLANT COMMUNITIES

There is convincing evidence of interspecific competition among parasitoids, which stands in apparent contrast with examples from herbivorous insects. This is consistent with the predictions made on the basis of resource quality and patchiness. Such arguments can be extended by considering other examples of high-quality, patchily-distributed resources. In the cases of both dung and carrion, Chapter 1 presented strong evidence for intense interspecific competition. In both of these cases interference was a clearly demonstrated mechanism. Herbivorous insects may encounter similar situations: the location and exploitation of seeds (for example, Janzen, 1975). There is also one stage, reproduction, when all male insects, whatever their food supply, must locate a patchy but high-quality resource female. If the prediction holds, then we would expect to find intense intraspecific competition for mates. Sperm competition illustrated this (Chapter 2), and two recent reviews (Baker, 1983; Fitzpatrick and Wellington, 1983) have shown that conflict and territoriality are associated with mating in a broad range of insect groups. Figure 7.3 summarizes the probable variation in the intensity of interspecific competition in some major feeding groups of insects.

As in the case of plants, we again end up with the paradox of higher competition where resources are of higher quality. The resolution here may be that low-quality resources produce exploitation competition whereas high-

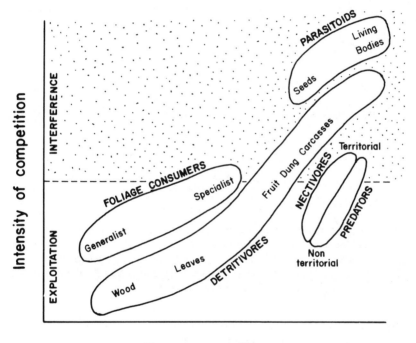

Resource quality
(Resource patchiness)

Figure 7.3 Postulated relationship between the intensity of competition and resource quality for major feeding groups of insects. Resource patchiness is correlated with resource quality, and is therefore included on the horizontal axis. This illustrates how studies of phytophagous insects may be unlikely to detect competition, whereas studies of insects feeding on corpses and dung are likely to detect it. The shaded region indicates where interference competition may be expected to be important.

quality resources produce interference competition. Therefore, when making general predictions about competition it may be necessary to separate these two kinds of competition. We are still left with a question, however. If resources are of low quality, will competition be intense because of resource limitation, or will be it be weak because the abundance of organisms is largely controlled by the adversity of their environment? This suggests a convergence with the debate regarding resources and competition in plants. Are herbivorous insects feeding upon widespread but well-defended plants (apparent plants *sensu* Feeny, 1976; stress-tolerant plants *sensu* Coley *et al.*, 1985 and Southwood *et al.*, 1986) the equivalent of plants occupying habitats with continuously low resource levels? If so, what can we say about the intensity of competition under these circumstances? Also, what are the consequences for other groups of organisms occupying habitats with low resource levels?

Figure 7.4 summarizes this possible convergence between plants and animals. The first (vertical) axis of the model represents frequency of

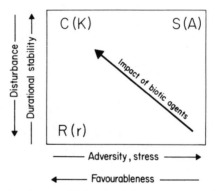

Figure 7.4 Two major forces which may influence community structure and the evolution of life-history traits: disturbance and stress (or adversity). The letters outside the parentheses follow Grime (1974, 1977): R, ruderals; C, competitors; S, stress tolerators. The letters inside the parentheses follow Southwood (1987, 1988): r, r-selection; K, K-selection; A, adversity selection.

disturbance, or its inverse, durational stability. The second (horizontal) axis represents stress or adversity, and their inverse, habitat favourableness. The way in which traits of organisms should vary along these axes has been discussed by Grime (1977, 1979) and Southwood (1987, 1988). The diagonal line represents their prediction about how the intensity of biotic interactions such as competition should vary among habitats. The model has been erected largely to explore the evolution of different sets of life-history traits in organisms, but it may also be an appropriate starting point for exploring how competition organizes communities. Is it possible to draw isopleths for competition intensity? Are asymmetric interactions predominant in one region? Is one region conducive to resource partitioning and another to dominance control? Do mechanisms of coexistence differ among regions? These and other questions remain unanswered. The next sections explore how such questions might be posed and answered.

7.4 MAKING THEORIES OPERATIONAL FOR HYPOTHESIS TESTING

A theme throughout this chapter and many of the above references has been that some environments are inherently unsuitable for organisms, and that this has consequences for the intensity of competition in those communities. A plethora of terms exists to describe those habitats thought to be somehow unsuitable for specific organisms: these include 'harsh' (McNaughton and Wolf, 1970; Schoener, 1986), 'physically controlled' (Sanders, 1968), 'physiologically less congenial regions' (Connell, 1972), 'unfavourable' areas (Southwood, 1977), 'stressed' habitats (Grime, 1979) or 'adverse' situations (Greenslade, 1983; Southwood, 1988). The difficulty in specifying units of measurement for these terms leaves them open to the criticism of Harper (1982) that they are '... little more than the observer judging what I don't think I'd like if I was a buttercup, kangaroo, flea, beetle, etc.'. This problem has three serious consequences for the development of population and community ecology:

1. We cannot synthesize the existing literature on environmental gradients since we do not know how to compare the habitats being investigated;
2. We cannot move towards predictive ecology (*sensu* R. H. Peters, 1980a) since we do not know how to describe the degree of unsuitability of a habitat to predict a particular ecological trait;
3. We cannot test many published hypotheses which use this non-operational vocabulary.

One solution to these problems might be to measure many environmental factors in each investigated habitat, particularly as automated recording devices can now measure many factors simultaneously and repeatedly. However, the weakness of this approach was emphasized more than 50 years ago by Clements (1935), who argued that analysis of habitats by physical factors alone is inadequate and that it is necessary to express physical and chemical measures in terms of plant functions (see also Weaver and Clements 1929). He suggested that '... community phytometers are often desirable and these range from sod cores and sown and planted quadrats to closures of several sorts'.

Although contemporary theorists have not always emphasized operational measures of unsuitability, Clements' suggestion appears to be the obvious way of making these concepts operational and therefore falsifiable. Imagine, in the simplest case, the comparison of two habitats. Select one or more species, and introduce propagules, juveniles or adults to both habitats. To measure the direct effects of the environment upon these individuals, remove neighbours which could be competitors, and exclude predators. The introduced individuals occupy the clearings for a specified period. The performance of these transplanted individuals could be measured in many ways, depending upon the interests of the investigator, but traits such as weight gain and reproductive output are obvious possibilities. Unsuitability is then measured by comparing performance; the greater the reduction in one plot relative to the other is, the more unsuitable the latter habitat. This is therefore a comparative measure of unsuitability, expressed in the units of performance of the actual organisms inhabiting the habitats of interest. The method can obviously be extended to include many habitat types, and as many species as necessary.

An experiment such as this can be regarded as a bioassay for measuring factors that are considered to be important in ecological theory. Although the following experiment illustrates the measurement of stress or adversity, it is easy to imagine extensions which could measure other important environmental variables such as disturbance and grazing. Weldon and Slauson (1986) have made analagous proposals, but they suggested using physiological measurements upon individuals already present. This has the advantage of avoiding the need to transplant phytometers, but it has disadvantages if the same species is not present in all habitats which are being compared.

7.4.1   An example. Comparing stress in two plant communities

Gradients of exposure to waves are a common feature of both marine and freshwater shorelines. On lakeshores exposure to waves not only produces a disturbance gradient but, since the substrate is physically sorted, a fertility gradient is also produced. This gradient stretches from infertile, sandy, wave-washed beaches to fertile, organic, sheltered bays (Keddy, 1983; S. D. Wilson and Keddy, 1985). To test whether this creates a stress gradient, and measure the intensity of the stress, ramets of a common shoreline rush were transplanted to the two ends of this gradient. To measure stress differences, some transplanting sites were cleared of all other plants. To compare the intensity of competition with the effects of stress, replicates were also set up where the ramets were transplanted into established vegetation.

Figure 7.5 shows that in the cleared plots there were dramatic differences between the two sites. Performance was reduced by 70% on sand shores. The performance of plants on sandy beaches grown with neighbours and without neighbours is not that different; compared with growth in clearings in bays, a full 90% of the reduction on sand beaches is due to abiotic effects, with competition reducing performance only a further 10%. Thus, on sand beaches the overwhelming constraint on performance is stress. However, in sheltered bays the difference in performance between cleared and uncleared plots shows that low performance in established vegetation is primarily due to the presence of neighbours. Not only do these two sites have very different levels of stress, but there is a clear inverse relationship between stress and competition.

Experiments such as these may provide the means to escape the proliferation of untestable, non-operational theories. They will require ecologists to place

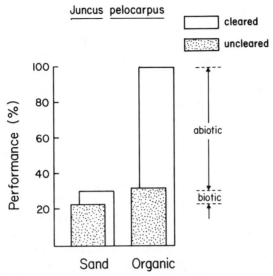

Figure 7.5 Making theory operational: measuring stress (adversity) in two contrasting habitats, using transplanted individuals. This example used the rush (*Juncus pelocarpus*) as a 'phytometer' to compare shoreline habitats. The cleared plots had all neighbours removed to measure the response of the rush to the environment independently of the effects of neighbours. Uncleared plots with neighbours present are also shown to illustrate the effects of competition intensity (S. D. Wilson and P. A. Keddy, unpubl. data). The arrows on the right show that a majority of the difference in performance between organic and sand habitats is a consequence of stress (adversity), and that competition intensity is higher in the organic site.

increasing emphasis upon carefully designed field experiments, and put considerable thought into to best 'organisometers' (with apologies to Clements). We will then be able to test the general principles regarding habitat unsuitability for which ecologists have been searching. In the absence of such operational measures of falsifiable predictions, ecological theory is reduced to rhetoric and scholasticism (Stearns, 1976; R. H. Peters, 1980a; Keddy, 1987).

7.5 RESOURCE PARTITIONING REVISITED

The search for environments varying in competition intensity provides a tool for extracting general principles from a wealth of special cases. An underlying assumption of this chapter has been that these environments can be recognized from external constraints placed upon the organisms. These external constraints included disturbance, stress, and variation in the distribution and quality of food. There is at least one intrinsic trait of communities which could similarly determine competition intensity. Therefore, predictions about the importance of competition based solely upon extrinsic factors could be misleading. The model of resource partitioning presented in Chapter 3 assumes that the more differentiation there is in fundamental niches, the less intense competition will be (for example, MacArthur, 1972; Vandermeer, 1972; Cody, 1974; Whittaker and Levin, 1975; May, 1981; Pianka, 1981, 1983; Giller, 1984). One could generate a contrary prediction by arguing that those sites where most intense competition occurred in the past are exactly those where niche differentiation should now be highest, and therefore interspecific competition lowest. One could therefore predict that competition should now be most intense in those sites where it has historically been least intense, because these sites are the ones where niche differentiation does not reduce interspecific competition.

A further complication is the prediction that increased interspecific competition may lead to convergence rather than niche differentiation. Aarssen and Turkington (1985) observe that although the assumption that interspecific competition leads to niche differentiation has become axiomatic in ecological theory, there is evidence for the opposite view. When comparing competitive interactions among plants in pastures of three ages (2, 21 and 40 years), they found more evidence for convergence in competitive abilities than for niche differentiation (see also Fig. 6.5).

Figure 7.6 illustrates these conceptual and methodological issues raised by resource partitioning. Past ideas have postulated the existence of and stressed the importance of arrow c – increased competition intensity leading to increased resource partitioning. Through feedback loop b, competition intensity is reduced and all species continue to coexist. However, there is an

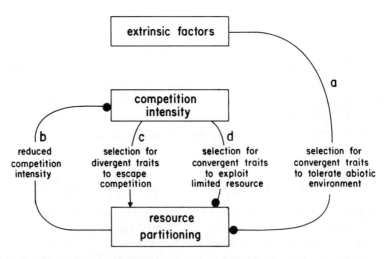

Figure 7.6 Evolutionary feedback loops between competition intensity and resource partitioning. There may be selection for increased partitioning by path c or decreased resource partitioning by paths a and d. In the absence of specific tests, the expected relationship (if any) between resource partitioning and competition gradients is a matter of speculation.

opposite effect, as shown by arrow d; there could also be selection for equivalence as all species are selected for increased competitive ability for the same limiting resource. In this case there is decreased resource partitioning and therefore a net positive-feedback loop (d + b) which increases competition intensity; there may still be coexistence because of increased competitive equivalence (as discussed in Chapter 6). Thus, depending upon which feedback loop one emphasizes, exactly opposite effects are predicted. In addition to all of this, the direct effects of extrinsic factors (arrow a) select for reduced resource partitioning by selecting for convergent traits to tolerate the abiotic conditions! It is by no means clear whether convergence or divergence should be expected; neither is it clear whether these situations can be detected or measured.

Differences in niche differentiation could therefore confound studies which compare sites varying in extrinsic controls. If niche differentiation were operationally defined, then a variety of questions at least peripherally related to competition could be posed. Does niche differentiation vary among different kinds of environments? If so, does niche differentiation tend to increase or decrease with disturbance, stress, food quality, etc.? How is this related to competition gradients? If, say, opportunities for niche differentiation decrease as extrinsic factors lead to increasing competition intensity, then extrinsic and intrinsic factors could act together to increase the steepness of the competition intensity gradients.

This suggests another way of comparing communities in order to look for constraints upon competition: the comparison of intrinsic as opposed to extrinsic factors (Table 7.3). This requires recognition of the distinction between resource competition and space competition (Chapters 1 and 3). Most studies and theory fall into the first column. This book, especially Chapter 6, places increased emphasis upon the second. The two examples discussed in this chapter can be provisionally assigned to these columns, with motile animals in the first, and plants and sessile animals in the second. The third column probably applies to a subset of communities structured by space competition. Underwood and Denley (1984) have emphasized the importance of founder control ('pre-emptive competition') in intertidal communities; Platt and Weis (1985) and Grace (1987) have provided recent examples for plant communities.

Schoener (1986) has taken this process in a different direction by producing a classification of community types including six organismic (intrinsic) and six environmental (extrinsic) axes. Surprisingly, stress ('severity of physical factors') is included, but there is no disturbance axis except for one labelled 'long term climatic variation'. Resource partitioning is included, but as an extrinsic axis ('partitionability of resources') rather than as an intrinsic one. The actual measurement of these axes is quite another problem.

The comparison of different kinds of communities is obviously a tool with a lot of potential. It is essential to emphasize again the importance of operational measures and testable hypotheses in this process. In this case

Table 7.3 Three possible kinds of community structure produced by competition. This classifies kinds of communities by their biological structure (intrinsic attributes) rather than by the environmental constraints upon them (extrinsic controls). Classification terminology from Yodzis (1978, 1986)

	Basis of competition		
	Resources	*Space*	
	Niche control	*Dominance control*	*Founder control*
Processes	Resource partitioning	Competitive hierarchies	Dispersal
	Symmetric competition	Asymmetric competition	Colonization
	Different fundamental niches	Inclusive fundamental niches	Gap-creation
Questions	Number of niche axes?	Traits predicting dominance?	Traits for dispersal and colonization?
	Coexistence and limiting similarity?	Coexistence through competitive equivalence?	Sensitivity to disturbances which create gaps?
		Variation in hierarchy position with environment?	
			Differentiation of regeneration niches?

simple measures of the degree to which different communities are niche differentiated will be essential for the application of empirical approaches.

7.6 TOWARDS GENERAL PRINCIPLES

When seeking generalizations about the role of competition in nature, it is vital to draw a distinction between 'competition intensity' and 'the importance of competition' in communities (Weldon and Slauson, 1986). Competition intensity is a measurable effect of the impact of neighbours upon the performance of individuals. In contrast, importance compares reduction in performance attributable to competition to that attributable to all other environmental factors. It is therefore possible to find that competition intensity causes a significant reduction in performance in a specific habitat, but that it is relatively unimportant when scaled relative to other factors such as environmental stress or predation. Actually measuring importance isn't as easy as measuring intensity because of the large number of other factors which need to be measured to compare with effects (if any) of competition. However, proposals by Weldon and Slauson (1986) and the analogous 'phytometer' approach advocated in this chapter do make measurements of importance possible.

This chapter has emphasized the role of comparative studies in exploring competition and detecting general patterns in nature. R. H. Peters (1980b) has drawn attention to the distinction between natural history and ecology: the former is the collection of facts like insects pinned in giant drawers; the latter is the search for general principles. These are actually two very different world views, and the search for general principles is not always appreciated. Some of the difficulty seems to lie in the degree of generality with which particular individuals feel comfortable. Part of this is simply human nature and has nothing to do with the validity of scientific generalizations. What for some is trivial detail, for others is conceptual richness. When friends return from holidays with five carousels of slides, their ability to provide details of their trip is usually more than their audience can appreciate. The audience's request for a few general principles is usually treated as a symptom of their vulgar inability to appreciate the higher levels of human experience.

Beyond this aspect of human nature, different levels of generality are necessary for different ecological scales (Starfield and Bleloch, 1986; Allen and Starr, 1982), and ecological theory will probably eventually consist of a series of nested models. The specific models needed for management of individual systems will use site-specific information and species nomenclature; these models will be nested within more-general models dealing with relationships among state variables and functional groups of organisms (Day et al., 1988). By applying a range of models to a system, one could then explore a situation from different hierarchical perspectives, ranging from precise predictions about one endangered species at a site to those about entire functional groups or guilds, under different management regimes (for example, van der Valk, 1981; Severinghaus, 1981; Day et al., 1988). A general description of pattern and process in riverine marshes may appear gross negligence to the dedicated naturalist, yet overwhelming in detail to someone who is interested in sessile organisms. For example, Grubb (1985) criticizes Grime's (1977, 1979) generalizations regarding evolutionary strategies by showing the rich detail of sub-strategies possible within each of the main strategies. This is offered as a criticism of Grime's work, rather than for what it really is – an elaboration of the richness of natural historical detail which is lost in any general model, and a reminder of the need for models with different degrees of generality.

However, generality does not mean the use of vague, poorly defined and non-operational axes. R. H. Peters (1980a) and Rigler (1982) have argued that empirical models are a powerful and rigorous tool for finding general principles. In contrast, *explanations* of phenomena cannot be evaluated except by how they either satisfy or entertain ecologists. Neither satisfaction nor entertainment appear to be worthy or objective criteria for evaluating scientific theories. In contrast, simple regression models provide clear criteria for testing the strength of relationships among variables and, by their very nature, force ecologists to use measurable entities in their models. A criticism of this approach as it is advocated to R. H. Peters (1980a) and Rigler (1982) is its failure to incorporate experimentation. This is something that could easily be rectified by designing appropriate bioassay experiments as described above (see also S. D. Wilson and Keddy, 1986a; Gaudet and Keddy, 1988).

One of the most productive avenues for future research in competition may be the combination of comparative studies, bioassay-type field experiments and empirical ecology. In the absence of such rigour, the value of comparison is limited. In its absence: (1) we cannot test existing published hypotheses which use non-operational axes; (2) we cannot synthesize the existing literature on environmental gradients, since we do not know how to compare the habitats being investigated; and (3) we cannot make testable predictions, since we do not know how to describe habitats to predict a particular ecological trait. These problems can be surmounted by the combination of comparison, experimentation and empiricism. This provides many unexplored opportunities for real advances towards competition theory. In Chapter 8 we consider what the goal of such theory might be, and how the behaviour of ecologists themselves needs to be considered when planning research strategies.

7.7 QUESTIONS FOR DISCUSSION

1. How do we choose dependent and independent variables for the study of ecological systems?

2. What do you make of the paradox of resource limitation? Does competition intensity really increase with resource levels?

3. Why are parasitoids the size of Alien (that is, sizes greater than or equal to humans) not found on Earth? Is this a quirk of natural history, or something predictable from resource characteristics?

4. Consider the methods for classifying types of communities by both intrinsic and extrinsic traits. What are their relative merits?

5. Compare the simplicity of Grime's model with the complexity of Schoener's model. What are the consequences of such differences for developing ecological theory? Which, if any, of their axes are measurable?

6. Are operational definitions important? Why do journals continue to publish models where the authors fail to specify how the axes can be measured to test the models?

7. Why has disturbance received more attention than stress or adversity?

8. Compare and contrast natural history and ecology.

9. Consider the research directions suggested by Chapters 5–7. Which, if any, do you prefer, and why?

8 The path to competition theory

Most citizens despise the very idea of idiosyncratic
and personal self-expression... and expect the bitter
disciplines of adult life to stamp such tendencies out
if the schools fail to do so. And, of course, they are
right.

E. Z. Friedenberg (1976)

Chained through our attachments, we perceive the world
through our ideas, our thoughts, our mental constructs,
taking these concepts to be the reality itself.

J. Goldstein (1983)

Nothing is easier than leading the people on a leash.
I just hold up a dazzling campaign poster and they jump
through it.

J. Goebbels (in A. Rhodes, 1976)

A final chapter could have one of two purposes: to emphasize what is known or
to emphasize that which is unknown. Most is likely to be gained when we walk
to the edge of established knowledge and peer over. This may be particularly
important for a topic like competition, where it is easy to be buried in past
anecdotes, assumptions, theories, models, conjecture and data. Consequently
this chapter first considers what the goals of competition theory might be.

Assuming that we have established a research goal, how do we attain it?
Creativity is an essential element. Popper (1959) and Magee (1973) have
stressed the importance of new testable ideas – 'bold conjectures' – but where
do they come from? Are they really necessary? If so, how can we get more of
them? It is these bold conjectures which can create scientific revolutions (*sensu*
Kuhn, 1970), but the process by which they arrive is poorly understood. In fact,
it could be argued that because they are, so far as we understand, transrational
or irrational in their origin, they are not a part of science at all. Again we have a
paradox of sorts. On one hand, scientific progress depends upon creativity

and bold conjecture. On the other hand, the process underlying creativity does not appear to be logical, and therefore is not easily treated as a part of the scientific process.

Since bold conjecture itself may be beyond rational analysis, this chapter places particular emphasis upon the processes by which questions are chosen and answered. Choosing the questions to be asked is arguably the most important part of the scientific process, and, at least in theory, could be guided by sound rational criteria. By asking the wrong questions, it is unlikely that we will obtain the right answer. If we ask the wrong questions and use inappropriate tests, then progress is even less likely. The second part of this chapter therefore considers the process by which questions are asked and answered. The flow of topics follows the order of the scientific process: selection of question, selection of model system, selection of conceptual approach, publication.

8.1 GOALS FOR COMPETITION THEORY

Where do we want to be in 10 years' time, and how will we get there? In order to judge scientific progress, we must have an agreed-upon goal. I suggest that there are two fundamental goals for competition theory and for community ecology: the development of 'assembly rules' (*sensu* Diamond, 1975) and the development of 'response rules'. Although Diamond's (1975) methods for creating assembly rules have been thoroughly criticized (Connor and Simberloff, 1979), it is important to distinguish between his methods and his goals. We could describe the objectives of assembly rules as follows. Given (1) a species pool and (2) an environment, can we predict the abundance of the organisms actually found in that environment? This requires a series of measurements of inherent traits of species, and a measure of their connection to essential aspects of natural environments. Several examples currently exist. Van der Valk (1981), for example, divided wetland plants into life-history types and proposed a general model for how different water levels would produce different vegetation types. Haefner (1978, 1981) similarly developed a series of rules for predicting the species composition of foliage-gleaning passerine birds from the measured characteristics of forest. In the latter case the rules were very complex. Perhaps the degree of complexity was in part due to concentrating on a small group of very similar species within one functional group, whereas van der Valk (1981) used functional groups of very different kinds of species (annuals and perennials).

The second objective of competition theory and community ecology could be called 'response rules'. Given (1) a specified assemblage of species, (2) a total species pool, and (3) a specified perturbation, can we predict the composition of a future community? This would again require knowledge of key life-history traits in the species pool, and the way in which these interact with basic types of perturbations. This would require combining description with comparison with experimentation. Description would be necessary to delineate the species pool, to define the initial states of systems subject to perturbations and to describe naturally occurring states resulting from perturbation. Systematic comparison of the attributes of species (screening; for example, Grime and Hunt, 1975; Rorison et al., 1987; Gaudet and Keddy, 1988) would be necessary to compile the necessary ecological information on the species in the pool. Lastly, experiments would be necessary to determine which traits provide the capacity to predict responses to different kinds of perturbations.

The ultimate test of ecological theory (of which competition theory will be a subset) is whether ecologists can really say anything useful about the world. Those of us who have participated in environmental assessment hearings know the frustrating experience of often having little more than vague ecological generalities and natural history to offer to hearing boards. It is not surprising that scientific input is sometimes treated with low regard. The second goal requires ecologists to keep applications in mind continuously. We can develop all the elegant models we wish, live distinguished academic careers, publish numerous well-cited papers, and so on, but the ultimate test of the value of our work is whether we really can make predictions about the real world. How big do ecological reserves need to be to protect different groups of species? Which kinds of species will be the first to disappear due to toxic chemicals? How long will it take for coral reef communities to recover from siltation? Are there alternative stable points in boreal forest or tropical forest, and what are the implications for forest management? The big questions are all there (International Union for the Conservation of Nature and Natural Resources, 1980). It remains to be seen whether we can develop the models and theories to answer them. Kuhn (1970) notes that scientists tend to choose 'answerable questions' as opposed to 'important questions'. One hopes that these are not mutually exclusive categories. It is instructive to re-read Clement's 1935 paper on 'Experimental ecology in the public service' as a reference point and compare it with more recent activities in this area (for example, Usher, 1973; Holling, 1978; Barrett and Rosenberg, 1981; Mooney and Godron, 1983).

Although the 'progress of science' is a popular cliché, there is no general agreement on what constitutes progress or how to measure it. R. H. Peters (1980a) has explored this by dividing scientific ideas into two classes: theories and concepts. Theories are predictive or falsifiable statements about nature. They propose relationships between observable phenomena, and inform us of which events we are likely to encounter. Concepts, in contrast, are not falsifiable, although they may be part of every scientist's thinking. They provide a conceptual framework which helps to organize theories, and which may lead individuals to new creative insights. Peters argues that ecology places far too much emphasis on concepts rather than theories, with the result that 'we have become modern scholastics interminably discussing questions which cannot be solved or tested scientifically'.

How do we measure the value of different kinds of questions and different kinds of approaches? This will depend upon the relative emphasis that we place upon theories and concepts. If the construction of theory is our only objective, then published studies are useful only to the degree that they allow prediction of patterns in nature. Statements such as 'competition intensity (the dependent variable) increases with the biomass of the community (the independent variable)' would be the sort for which we would be testing. Alternatively, concepts have utility if we see science as an activity which expands the horizons of human experience. In this case we can be satisfied if we have increased our 'understanding' of nature. We might, however, be further challenged to provide an actual measure for 'understanding', or to prove that increasing our personal understanding leads to an increase in other people's understanding. Much of the sterile debate within the scientific community occurs either between individuals who have different ideas about utility or between individuals who have different concepts. In either case no amount of data will resolve the situation.

Judging the value of different research goals and methodologies also requires consideration of how scientific progress actually occurs. There is no consensus on this point. Contemporary views about how science is actually practised fall along a continuum. At one end data and facts are everything, at the other end they are unimportant and are collected only to amplify belief systems. At least five positions along this continuum can be recognized.

1. Science primarily involves the patient collection of facts. They are important for their own sake. There may also be the belief that someone (but not the collector) will eventually make sense of them through induction. This is data for the sake of data, or natural history (*sensu* R. H. Peters 1980b).

2. Data are important for falsifying hypotheses, but 'bold conjecture' or original hypotheses are what drive scientific progress. Data are collected in order to falsify hypotheses, and their collection is guided by the question being asked (Popper, 1959; Magee, 1973).

3. Data are collected to solve small technical problems, but there is a larger context or paradigm shared by scientists. The data are collected to clarify aspects of the paradigm, but not to challenge it (Kuhn, 1970).

4. Science is primarily political. Individuals of perceived high status dictate the prevailing world view. Data are collected to verify this world view. Contrary data are not collected and, if they are, are not accepted (Fagerstrom, 1987) and cannot be published (Mahoney, 1976).

5. Science is part of the entertainment industry, and the objective of scientific papers is to tell entertaining (i.e. intricate and elaborate) stories to a well-educated audience. Short papers which pose a clear question and provide an answer are either 'naive' or 'least publishable units' (LPUs).

This chapter (and this book) assumes a position between regions 2 and 3 of the continuum, and treats the existence of regions 1, 4 and 5 as obstacles to scientific progress.

8.2 BRAINS AND THEIR LIMITATIONS

Any discussion of science and scientific progress has to begin with the arena itself: the human brain. Anything short of an entire book on cognitive processes is necessarily superficial. However, the issue is raised here because it is an important (and often neglected) reference point for the sections that follow. As Loehle (1987) observes: 'problem solving is a mental activity, subject to psychological biases and shortcomings, rather than an exercise in pure logic'. If this is true for a prescribed process like problem solving, then we may expect it to be even more important for subjective processes like the choice of problem to be solved.

When scientific research is conducted and models are constructed, the human mind is usually pictured as an objective observer of some external reality. At the same time scientists believe that the human mind evolved from simpler nervous systems. We thus have a paradox: on the one hand, we are dispassionate observers, on the other, active participants.

There surely is no biological option but to adopt the latter view: our brain as a part of nature, with its own constraints and peculiarities of natural history. Dawkins (1976) presents a compelling picture of the evolution of human consciousness. He describes living orgnisms as survival machines built by genes. Traits ranging from increasing cell size, through multicellularity to consciousness can all be explained as traits which enhance the stability of survival machines in unpredictable environments. Consciousness can be considered to be the ability to simulate. By simulating possible future situations, an organism is able to decide which of n possible situations appears to be most desirable. Simultaneously, it is able to practise responding to situations which it has not yet encountered. Dawkins concludes that whatever the philosophical problems raised by consciousness, from an evolutionary perspective it can be thought of as the culmination of an evolutionary trend for genes to put nervous systems in charge of day-to-day decision making. It is this nervous system which we, as scientists, use to explore reality.

There are two main constraints on the exploration of nature imposed by our minds: constraints which are inherent in the organization of human brains (hardware constraints) and constraints which are learned (software constraints). These constraints have been explored in detail elsewhere (for example, Mahoney, 1976; E. O. Wilson, 1978; Hofstadter and Dennett, 1981; Bateson, 1979; Hayward, 1987; Loehle, 1987). They operate during two mental activities which are products of our evolutionary heritage: perception (sensory input) and interpretation of these perceptions (information processing).

The process by which these inherited and learned behaviours can influence apparently logical thoughts and consciousness is receiving increasing study.

Two contrasting sources of evidence suggest that by the time conscious thought arises, emotions, urges and tendencies have already shaped it. Hayward (1987) has explored two different traditions which provide data on this process. The first is meditation, the objective of which is to bring previously unconscious mental activities into awareness so that such mental activities can be observed while they are happening. When this training occurs, it is observed that before each conscious thought there is a rapid series of steps where emotions and habitual patterns determine which conscious thought will arise. Hayward then explores recent data from the cognitive sciences, which explore the processes of human perception. They yield similar conclusions: preconscious tendencies determine which conscious thoughts will arise. Conscious thinking is only a small buoy floating on an ocean of unconscious mental activity.

This has direct consequences for research. Precisely because emotions, urges and habitual patterns occur before consciousness arises, conscious thinking is unable to deal with them. The mind can continue to look for perceptions which reinforce beliefs rather than challenge them ('confirmation bias'; Loehle, 1987). Humans are chained to habitual thought patterns precisely because they are normally unable to observe these habitual thought patterns and therefore cannot escape them. In this context, creative events probably occur when there is a gap in this flow of thoughts allowing a brief instant when something entirely new can enter – if the observer is in a state to detect and appreciate this new impression. Unfortunately, the very momentum of conditioned habitual thoughts makes it unlikely that insights arising in such gaps will be noticed and acted upon. Fortunately, this analysis also suggests that increased awareness of habitual thoughts can simultaneously allow escape from habitual thought patterns and openings to creative instants. Such issues are discussed further in books such as Jung (1964), Trungpa (1976), Hofstadter and Dennett (1981), Bateson (1979) and Hayward (1987). Is there any evidence that such issues are relevant to progress in the study of competition? The following sections explore this possibility.

8.3 CHOOSING A QUESTION

8.3.1 The myth of objectivity

The choice of question is the most important part of the scientific process (Chapter 4), but it is also the most subjective aspect of the scientific process. What is perceived to be important can be influenced by a wide variety of cultural factors (Fig. 8.1). Hypothesis testing – the middle part of the figure – is where scientists are most able to escape psychological bias, because there are specified rules to follow. The actual testing of hypotheses is limited only by honesty, perception, knowledge of taxonomy and statistics, budgets and awareness of unexpected confounding variables. The more we move away from this narrow region of formalization, the more conscious and unconscious

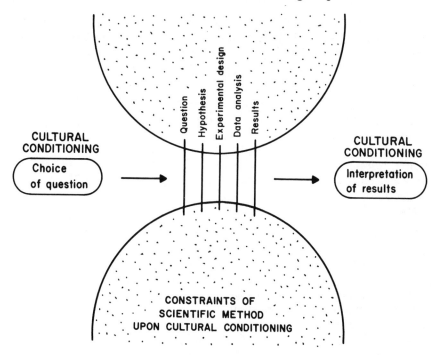

Figure 8.1 Influence of cultural conditioning upon science. Cultural influences are greatest when questions are chosen and results interpreted. The less awareness there is of cultural conditioning, the greater the impact that it will have upon this process.

biases can act. They are probably most critical in the choice of new questions and the interpretation of tests. When we consider the effects of the culture in which scientists operate, it is important to separate the effects of culture as a whole from the effects of the scientific culture, as these may act in different ways and at different scales.

8.3.2 Socio-economic culture: mutualism and the communist conspiracy

There are an infinite number of questions to be answered in science. Why study competition, and why write a book about it? Presumably because it is (or at least has the potential to be) an important force organizing the biosphere. Table 8.1 shows the impression transmitted to students regarding the relative importance of competition, predation and mutualism in organizing the biosphere. Table 8.1 suggests that mutualism is comparatively unimportant compared with competition.

Consider – each cell in our body is possibly a symbiotic association of prokaryotes (Margulis, 1970). A large proportion of the world's biota is made

Table 8.1 The impressions given to students regarding the importance of the three major ecological interactions in the biosphere, as assessed by the number of pages on the topic referred to in the index of current textbooks on introductory ecology

Textbook	Mutualism	Competition	Predation
Colinvaux (1986)	1	33	70
Collier et al. (1973)	0 (1)[a]	45	30
Hutchinson (1978)	0 (9)	59	6
Krebs (1978)	3	50	32
Lederer (1984)	5	21	4
McNaughton and Wolf (1979)	20	77	71
Odum (1983)	15	17	15
Pianka (1983)	3	74	41
Ricklefs (1979)	3	38	30
Ricklefs (1983)	2	11	14
Smith (1986)	2[b] (1)	19	24
Whittaker (1975)	5 (9)	18	22

[a] The number in parentheses is symbiosis which some authors equate with mutualism.
[b] Mutualism not in index, but present in text.

up of multicellular organisms – a mutualistic association of unicellular components (but see Buss, 1981, 1988). We all breathe oxygen that is produced by plants. Studies of mycorrhizae suggest that plants are joined by extensive mycorrhizal networks. All multicellular organisms have endosymbionts in their guts, which may assist with digestion and/or manufacture vitamins. Many plants require insects for pollination and seed dispersal. Predators kill herbivores which otherwise would eat plants. (At this point the author has concluded that he should be writing a book on mutualism, and readers have probably concluded that that is what they should be reading.) The data are clear (see also Boucher *et al.*, 1982). Yet the view apparently shared by writers of ecology textbooks, and the impression given to our students, is that mutualism is relatively unimportant. Why?

I offer five hypotheses. All centre on the idea that scientists consciously and subconsciously draw upon their culture for models. Toffler (1984), for example, observed that during the machine age scientists tended to generate machine-like models of nature. Prigogine and Stengers (1984) state '... many scientific hypotheses, theories, metaphors and models (not to mention the choices made by scientists either to study or to ignore various problems) are shaped by economic, cultural, and political forces operating outside the laboratory'. This is in a sense obvious, because scientists have few other options except perhaps revelation, and even that must be interpreted by the receiver in terms with which he or she is familiar.

Culture

Scientists choose their models from the set of concepts available to them; they cannot use concepts of which they are unaware. Ecology has developed rapidly in the USA. Therefore, we need to look critically at the USA to see whether the prevailing socio-economic environment may have influenced research strategies and conceptual frameworks. One perspective, among many more-positive ones that could be offered, emphasizes the degree to which competition controls US foreign and domestic policy. The USA is a capitalist country with a large economic empire. Its economy is based upon concepts such as exploiting resources, competition between companies and corporate takeovers. Resources are extracted from Third World nations (Lappe and Collins, 1982; Myers, 1985), and military aid is given to install compliant dictatorships whose death squads remove those who propose alternative models to capitalism (Chomsky and Herman, 1979; Klare and Arnson, 1981). There is institutional intolerance of alternative proposals for the organization of American society: in the 1950s McCarthyism purged American universities of individuals who taught alternatives to capitalism. Even today there is legislation to stop 'subversives' from visiting the USA: the subversive category has included prominent Canadian authors such as Farley Mowatt. This is not a popular view, and Chomsky (1987) has documented the degree to which the North American media prevent such views from being made accessible to the public. Scientists can only draw models from the possibilities of which they are aware, and perhaps ecology has been hampered by restricted access to individuals (and ideas) offering co-operative models for society and nature.

Excitement

Another hypothesis for such bias is the human search for excitement. Watching something kill something else is exciting, and horrific if you identify with the victim. There is a class of nature films and children's storybook which has this view of nature: insect eats insect, frog eats insect, snake eats frog and hawk eats snake. Perhaps competition is considered exciting; this may be one explanation for the vast amounts of time invested in playing, watching and writing about sports. Perhaps scientists simply find co-operation boring.

Gender

Another hypothesis is that there is gender bias in ecology. Table 8.2 presents data on another topic, but equally provides a sample which suggests male dominance of ecology. Although gender-linked behavioural differences are still both poorly understood and controversial (for example, Maccoby and Jacklin, 1974; Gould, 1981; Lowe, 1981; Rothschild, 1983), the extensive review of published studies by Maccoby and Jacklin concluded that levels of

Table 8.2 Community structure of community ecologists assessed by invitations to write chapters, the number and length of contributions, and citations. The three volumes used were: 1, Cody and Diamond (1975); 2, Strong *et al.* (1985); and 3, Diamond and Case (1986)

Author[a]	Country	Taxon	chapters[b]			Pages[c]			Citations[d]		
			1	*2*	*3*	*1*	*2*	*3*	*1*	*2*	*3*
Addicott, J. F.	USA, Canada	Insects	0	0	2	0	0	21	0	1	8
Brown, J. H.	USA	Mammals	1	1	1	27	15	21	5	15	21
Case, T. J.	USA	Lizards	0	0	3	0	0	35	0	4	10
Chesson, P. L.	USA		0	0	2	0	0	28	0	0	9
Cody, M. L.	USA	Birds	1 (1)	0	1	56	0	25	8	14	15
Colwell, R. K.	USA	Birds	0	1	1	0	16	19	3	5	17
Diamond, J. M.	USA	Birds	1 (1)	1	5	115	29	83	13	14	28
Gilpin, M. E.	USA	Birds	0	1	1	0	29	18	3	7	18
Grant, P. R.	USA	Birds	0	1	1	0	31	19	4	29	41
James, F. C.	USA	Birds	1	1	0	34	20	0	0	5	2
May, R. M.	USA		1	(1)	0	40	14	0	7	11	32
Schoener, T. W.	USA	Birds	0	1	2	0	28	44	10	11	40
Simberloff, D. S.	USA	Birds	0	2	0	0	39	0	3	24	15
Werner, E. E.	USA	Fish	0	1	1	0	23	15	1	11	12
Wiens, J. A.	USA	Birds	0	1	2	0	19	28	0	19	22

[a] Authors must have contributed at least two chapters among the three texts to be included in the table.
[b] Introductions authored are shown in parentheses.
[c] Includes authorship of introductions and rejoinders.
[d] Number of the author's papers cited in the book. Note that this is a conservative measure of the number of times that the author's work is cited.

aggression differ between genders. They cited four sources of evidence: (1) males are more aggressive than females in all human societies for which evidence is available; (2) behavioural differences arise early in life when there is little evidence for parental moulding of aggression levels; (3) similar differences occur in subhuman primates; and (4) aggression is related to levels of sexual hormones and can be manipulated by experimentally modifying levels of these hormones. Maccoby and Jacklin stress that there are many unresolved issues in such research, but we may speculate that if females had dominated the upper echelons of ecology, nurturing co-operative interactions would have received the emphasis it deserves.

Taxonomic bias

A following section on choices of model systems shows that ecological research is conducted upon a highly atypical group of organisms, birds and mammals, which do not represent the actual composition of the biosphere in number of species, biomass or life-cycles. It may be that these atypical members of the biosphere are particularly influenced by competition and predation, but that if more-representative organisms had been chosen, mutualism would have been considered typical. The problems of taxonomic bias are discussed in more detail below.

Scientific community structure

Within the scientific system itself there is competition for limited research funding and competition for space in journals. Within departments there is competition for space and equipment. Mahoney (1976) and Sindermann (1982) have provided different perspectives on these aspects of the psychology of scientists.

8.3.3 Elites and the exercise of power

Another factor which influences the choice of questions is the tendency for a few individuals to set an agenda which is then followed by others. The agenda varies from year to year, but topics such as 'resource partitioning', 'competitive exclusion', 'null models' and 'heterogeneous environments' rise to prominence and then are eclipsed by a new fashion. A restricted group of scientists ride into prominence on the crest of each such wave. How might we identify and describe the group which sets the agenda for others to follow? As one possible route, Table 8.2 lists authors who have repeatedly contributed to volumes on community ecology. The selection of an agenda for ecologists appears to lie in the hands of relatively few individuals. As the next section shows, at very least this group shares a strong bias in selection of model systems.

Dye and Zeigler (1987) have explored and analysed the role of élites in the USA in some detail. They are particularly interested in the way in which such élites exercise power. They observe that although élites may have the power to make policy decisions directly, equally important is the power of deciding what issues will be discussed. 'Power is not only deciding the issues but also deciding what the issues will be' (Dye and Zeigler, 1987, p. 94). This form of power can be exercised in three ways: (1) élites may act directly to exclude an issue from discussion; (2) subordinants may anticipate the negative reaction of élites and ignore proposals or suggestions that would disturb the élite; or (3) the underlying values of society itself may prevent serious consideration of alternative programmes and policies. Under such circumstances the top institutional structures need not exercise their power overtly, but allow it to be exercised by their subordinants. Interestingly, even the leadership of protest movements tends to come from within the élite. The degree to which the political and economic analysis of Dye and Zeigler applies to the structure of the scientific system is currently not clear, perhaps because scientists have received less attention. Attributes of the existing scientific system (e.g. élites and the anonymous peer review of grants and papers) suggest the possibility of strong institutional constraints on innovation and self-perpetuation of entrenched conceptual frameworks.

The implications of this for the behaviour of individual scientists are not clear – even Dye and Zeigler (1987) are unable to agree upon the consequences of their diagnosis. The tendency to create and follow leaders is not restricted to science, of course, so we may learn by considering an extreme case from another area of human interest where élites are well established and accepted. Consider the case of Krishnamurti, who was raised to be a messiah. He was born in India in the late 1800s, and was adopted by Annie Besant, president of the Theosophical Society in England. He was privately educated, and presented to the world as a 'World Teacher' who, according to various scriptures, appears from time to time to bring spiritual enlightenment to humans. One can imagine the response of his adherents (not to mention that of Annie Besant) when he renounced his role as the new messiah and dispersed his followers. Perhaps his message could apply to any system ruled by élites: '... leaders destroy the follower and the followers destroy the leader. Why should you have faith in anyone?'.

8.4 CHOOSING APPROPRIATE MODEL SYSTEMS

Once the important question is chosen, a model system is selected to answer it. The choice of system is of particular sociological interest. We shall ask three questions. What is the pool of systems available in the biosphere? How do perceived agenda-setters choose from this pool? How do scientists as a whole select from this pool?

8.4.1 Plants and animals: a non-existent dichotomy

In the Middle Ages it was widely perceived that some things were green and sedentary, and others were not green and moved. Clearly nature was made up of plants and animals. With increasing knowledge of the planet's biota, it is now recognized that a more realistic classification has at least five kingdoms: Monera, Protista, Animalia, Fungi and Plantae (Fig. 8.2).

Science has not really moved to incorporate this view. In ecology the terminology has been updated to 'autotrophs' and 'heterotrophs', but books are still written and courses offered on animal and plant ecology. Most universities still reveal their medieval roots by preserving departments of botany and zoology. To escape this medieval anachronism, we may ask ourselves: 'Would the principles of moneran ecology be different from those of fungi?' Do competition, predation and mutualism produce different kinds of structuring in communities consisting of different kingdoms or mixtures thereof?

These different groups of organisms probably differ in importance in the functioning of the biosphere, although it is not entirely clear how to measure

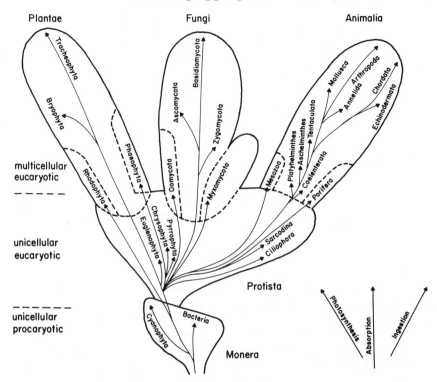

Figure 8.2 Plants and animals: a non-existent dichotomy. The five kingdom system of classification (Whittaker, 1969) recognizes three levels of organization: procaryotic unicellular (Monera), eucaryotic unicellular (Protista) and eucaryotic muticellular or multinucleate. There are three methods of resource acquisition: absorption, photosynthesis and ingestion. At the multicellular/multinucleate level these three methods of resource acquisition generate the final three kingdoms: Fungi, Plantae and Animalia.

importance. One approach might be to conduct a thought experiment by selectively removing different groups of organisms and estimating the perturbation which the biosphere would experience. A biosphere without nitrogen-fixing monerans, or a biosphere without decomposition by fungi would be vastly different from the biosphere of today. In contrast, a biosphere without birds would probably be rather more similar to existing conditions.

These groups of organisms also differ in abundance in the biosphere. Figure 8.3 shows relative abundances according to two criteria: numbers and biomass. The smaller groups of organisms are most poorly known, so the data on them are most difficult to tabulate. If Fig. 8.3 is contrasted with Table 8.2, then a striking anomaly is revealed. The choice of model systems in Table 8.2 bears no relationship to the biosphere, and could only be termed a clear and striking case of taxonomic myopia or ornithophilia.

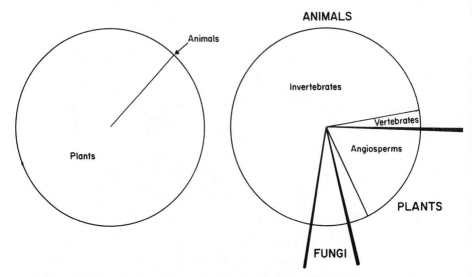

Figure 8.3 The importance of different life-forms in the biosphere, measured in two ways: biomass (left) and number of species (right). Biomass data are from Whittaker (1975). Species numbers were adapted from Colinvaux (1986, Table 1.2); fungi are almost certainly underestimated, whereas protists and monerans are excluded from the figure owing to lack of reliable figures. Note that all birds and mammals combined form only a subset of the small vertebrate slice.

8.4.2 The moose–goose syndrome

Of course, it may be that the general pool of practising ecologists is not influenced, consciously or subconsciously, by data such as those in Table 8.2. This hypothesis is testable. To explore the systems selected by ecologists as a whole, 167 papers were selected at random from 10 years of *Ecology* (1976–1985). Each paper was classified according to the system used and its geographic location. A null model of system selection would predict that systems should be selected in direct proportion to their abundance in the biosphere. Figure 8.3 presented this null model, both on a biomass basis and on a numbers basis. Figure 8.4 shows the systems used by ecologists who published in *Ecology*. Vertebrates are dramatically over-represented. There is no correspondence between the abundance of organisms in the biosphere and the effort invested in studying them.

Although the data are clear, hypotheses to account for them are less so. A proximate cause may be the influence of the authors in Table 8.2. This is not the only hypothesis, nor is it necessarily the ultimate cause, since it also raises the question of the origin of the patterns in Table 8.2. The following hypothesis is based upon approximately a decade of observations upon the natural history of biologists and biology students.

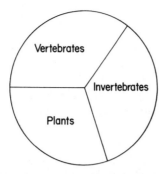

Figure 8.4 Taxonomic structure of ecological studies illustrated by the relative abundance of three taxonomic groups in 129 papers randomly selected from *Ecology* over the period 1976–1985 (a further 38 papers could not be assigned to a taxonomic category). Compare this figure with Fig. 8.3; which represents the null model for ecological studies. There is little correspondence between the abundance of organisms in the biosphere and the effort invested in studying them.

Much research is focused upon big things that are fun to kill, and things that fly and look pretty. (Things that the foregoing eat are admittedly also important, but generally can be lumped into categories of edible and non-edible.) University ecology programmes are often connected to wildlife biology programmes, at least in North America. Wildlife biology is a distinguishable social unit with recognizable dress code: work boots and cowboy boots, blue jeans, flannel shirts, four-wheel-drive trucks and gun racks. This syndrome is also associated with certain kinds of organisms, mainly big ones that are fun to kill and look good when parts are stuffed and hung on the wall, and for further information the reader is referred to men's outdoor magazines. The structure of biology departments can also reflect this bias, with the many zoologists hired to represent the diversity of the animals usually balanced by a couple of individuals covering the rest of the world's biota.

Ornithophilia is a special case of this syndrome. There are approximately 8600 species of birds in the world (Colinvaux, 1986) – probably less than the number of bird-watchers in the average English county. Birds are pretty, cute and visible, and listing is an excellent hobby. However, are they really an appropriate system for studying community ecology? In an apparently heretical article in a volume on ornithology, Karr (1983) observed that although the appeal of birds is widely known, there are certain problems in using them for research. Accurate censuses are very difficult. Relevant factors such as food availability are difficult to measure. Controlled experiments are difficult. At this point one might expect a recantation, but orthodoxy triumphs when, in the next paragraph, Karr begins '... I would not argue for abandonment of studies of avian communities...'. Of course, the point is not that birds should not be studied, but that organisms should be selected because they are good model systems for ecology rather than simply pleasant hobbies.

A review of themes in aquatic ecology (Khailov, 1986) reveals similar biases. In a sample of 8600 published papers, zoological studies of aquatic objects comprised 26% of the published papers compared with 6.5% for botany and a mere 0.8% for microbiology. (Two other aspects may also be relevant: studies of nutrition (68%) greatly outnumbered those of excretion (15%), even though in any open system inputs are equal in importance to outputs. Similarly, the growth of biomass was five times better studied than its decomposition (14% compared with 3% of published studies). It appears that the processes of excretion and decomposition also lack the attributes necessary to inspire ecological research.)

At this point we can only ask the question: 'What would a body of ecological theory look like if Plantae, Fungi, Monerans and Protistans played an appropriate role?'.

8.4.3 An historical tangent

Tansley's (1914) Presidential Address to the British Ecological Society dealt with many contemporary issues in competition, including the value of release experiments, the need for mechanistic studies, and the virtues of generality ('Quantitative results are of no use ... unless they have some kind of general validity ...'). Clements et al. (1929), in a book entitled *Plant Competition: An Analysis of Community Functions*, included a thorough review of competition concepts back to Malthus, an extensive series of transplant studies exploring competition, studies designed to explore competition for different limiting resources (light, water and nutrients), and a summary of the consequences of competition for community organization. The existence of a clear statement of conceptual approaches to studying competition (more than 70 years ago), followed by a synthesis (60 years ago) raises some important questions regarding the progress of science (see also Jackson, 1981). Given that there was and is overwhelming evidence for the existence of competition in plant communities, why did this early interest fail to stimulate a rigorous development of competition theory by plant ecologists? Why did it take so long for field experiments to become popularly accepted by plant ecologists (for example, Putwain and Harper, 1970; Sharitz and McCormick, 1973; Harper, 1977; Rabinowitz, 1978; Mueller-Dombois and Ellenberg, 1974; Grace and Wetzel, 1981; del Moral, 1983)? Why did plant ecology continue to be dominated by description? Why did none of the contributors to the three recent community ecology books (Cody and Diamond, 1975; Strong et al., 1984; Diamond and Case, 1986) cite Tansley (1914) or Clements et al. (1929)? If their work had served as a foundation instead of being overlooked, where would the study of competition be today? The answers to these question are important if we want to ensure that similar mistakes are not being made today.

8.4.4 Geographical bias

Geographical bias is a separate problem with separate causes. Ecological processes probably vary significantly around the planet. This would argue for spreading ecological research around the planet in a representative manner. Since there are many more species in the tropics, it would be reasonable to propose that ecological research should reflect the latitudinal diversity gradient, but Fig. 8.5 shows that research is clustered in the temperate zone. How far we can extrapolate from the temperate zone to the rest of the planet is not known. This figure is included so that it can be cited when writing tropical research proposals.

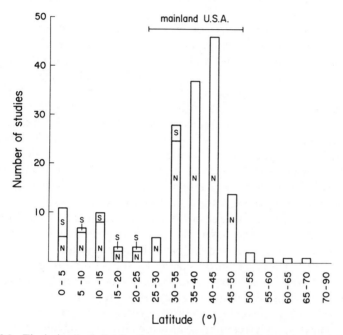

Figure 8.5   The latitudinal distribution of ecological research based upon 167 papers from *Ecology* randomly selected from the period 1976–1985. Given that both diversity and surface area per degree of latitude decline with latitude, the appropriate null models would have a maximum at the left and decline to the right.

8.5   SELECTING A CONCEPTUAL APPROACH

Having chosen a question and a model system, a conceptual approach must be selected. Chapter 4 discusses the range of conceptual approaches available for studying competition. Two issues are not covered in that chapter. Both of these issues relate to human behaviours used to explore nature. At this point it is not clear to what degree such behaviours are inherent mental tendencies and learned intellectual approaches. Whichever, they both have an influence upon problem solving.

8.5.1 Gaia and dualism (yin and yang or yin versus yang)

The human mind seems predisposed to seek dichotomies in nature: true believers and heretics; the saved and the damned; them and us; black and white. These and many other dichotomies are mental concepts which we construct and superimpose on a complex world. There are times when such classifications may be useful, but the examples above have been responsible for some of humanity's worst crimes. This suggests that dichotomies should be approached with caution.

In a world where we are raised on dichotomies, it is entirely understandable to carry this mental structure into ecology. It is less clear what the consequences have been and will be. However, it is obvious that most ecological questions are not dichotomous. Intuition alone tells us that it is naive to ask whether competition or predation structures communities; rather, we should be asking how they interact. This pluralistic view is becoming more widely accepted (Quinn and Dunham, 1983; Diamond and Case, 1986).

However, where in a culture built on dichotomies do we turn for intellectual models that escape from them? There have been several recent books drawing parallels between theoretical physics and 'Eastern philosophies' (for example, Capra, 1975; Zukav, 1979), suggesting, incorrectly in my view, that Western science is just rediscovering traditional Eastern world views. This superficial approach does, however, have elements of truth, in that different philosophical systems do provide a variety of world views and intellectual structures which may enhance our options for building ecological models.

Dayton (1979) has also made this point. He starts from Pirsig's (1974) book *Zen and the Art of Motorcycle Maintenance*, which Pirsig himself admits is authoritative on neither Zen nor motorcycles. More-authoritative approaches to Zen or Buddhist thought (for example, Watts, 1958) stress the same point, however: that questions set up to be answered by yes or no may in fact have a third logical alternative, *mu*. *Mu* means (roughly translated) that the context of the question is wrong and that it does not make sense to ask that particular yes or no question. Dayton concludes: 'Many of my own hypotheses were carefully designed to force yes or no answers from nature when, in fact, nature may have been crying out *mu*...'.

Levins and Lewontin (1982) have also raised several of these concerns, but their treatment had the stated objective of developing an implicitly Marxist approach to deal with them; although capitalism has undoubtedly distorted views of nature, it is not logical to assert that therefore the Marxist alternative must replace it. The capitalist–Marxist dichotomy is simply yet another dichotomy. A common misconception in students presented with dichotomies is that the truth is somewhere in between. However, this too is a trap because the dichotomy itself provides the context. The best social order for humans may be neither capitalism nor communism, but this does not mean it must be a mixture of the two. 'Neither left nor right but ahead' is the motto of the Green movement (Capra and Spretnak, 1984), which is branded as right by left and as left by right. We might wonder how ecology would develop if it asserted 'neither yes nor no but ahead'.

8.5.2 Beyond reductionism

One dichotomy which has influenced aproaches to the study of competition, and indeed the study of ecology, is the holism–reductionism dichotomy (Worster, 1977; McIntosh, 1985; Southwood, 1987). Interestingly, the holist–reductionist dichotomy is correlated with the experimental–descriptive dichotomy. It is probably accurate to say that plant ecology has, until recently, been dominated by descriptive holistic approaches (Harper, 1982), whereas in animal ecology experimental reductionist approaches have been more common (McIntosh, 1985). Plant ecology has gradually become more reductionist too, and undergraduates are still being taught how the holistic views of Clements were challenged by the reductionist views of Gleason, leading to the current burst of activity in plant population biology (for example, Harper, 1977; Solbrig *et al.*, 1979).

The reductionist view of nature is that the behaviour of complex ecological systems can be understood and eventually predicted only by breaking systems down into parts, studying the parts, and then reassembling them. The parts are normally assumed to be species, populations or individuals. This has led to a wealth of studies where individual species are selected and large amounts of information accumulated, usually to be published as a monograph. More recently, systematic pairwise removal experiments have been used (for example, Fowler, 1981; Silander and Antonovics, 1982) to assess the intensity of interaction among the many populations which comprise a community.

The holistic view of nature is that the behaviour of complex systems cannot be fully understood and predicted by studying pieces because there are 'emergent properties' or 'state variables' which are characteristic of assemblages but not of individuals. Although critics call this metaphysics or mysticism, Prigogine and Stengers (1984) have discussed the origin of organized 'dissipative structures' in non-equilibrium thermodynamic systems. The number of possible interactions among any group of components rises exponentially with the number of components, and these interactions can produce higher levels of organization. Since ecological systems are highly connected, the argument goes, it is highly improbable that the behaviour of communities will be predicted from studies of individual pieces.

Although the above are only caricatures of entire world views, they serve to present the basic dichotomy. Wimsatt (1982) provides one severe criticism of reductionism in his essay on reductionistic research strategies. Rigler (1982) provides another. In the latter case, Rigler discusses the problems with a reductionist research strategy in limnology:

> A temperate lake may support 1000 species. If each species interacted with every other species we would have $(1000 \times 999)/2$ or 0.5×10^6 potential interactions to investigate. Each potential interaction must be demonstrated to be insignificant or quantified. If we estimate one man-year per potential interaction it would take half a million man-years to gather the data required for one systems analysis model.

This may generate employment for ecologists (at least in the short term), but it seems unlikely to produce rapid scientific progress.

To avoid a dichotomy, the question might be phrased: 'Which ecological questions can be answered by reductionistic approaches, and which can be answered by holistic approaches?'. This would be a refreshing change from symposia with titles on the theme of reductionism versus holism. Instead of dichotomies, we may imagine a two-way ordination of ecological approaches (Fig. 8.6). 'Holistic experimentation' is probably the least explored approach to studying competition in nature (for example, Dayton, 1979; Sousa, 1984; Brown *et al.*, 1986). Its major traits are: (1) consideration of an entire community rather than a single population; (2) applying a perturbation in a field situation; (3) monitoring the responses of many dependent variables simultaneously; and (4) alternating frames of reference between individual components which are monitored (e.g. abundances of individual species) and emergent properties (e.g. richness and functional groups). The analogy with

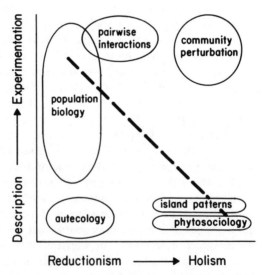

Figure 8.6 Some approaches to the study of competition in ecological communities. The broken line marks an axis along which dichotomies are frequently created.

physics and chemistry is the construction of ideal-gas laws rather than models based upon the activity and interactions of individual molecules (Prigogine and Stengers, 1984).

8.6 OBSTACLES TO COMMUNICATION

8.6.1 Peer review and status hierarchies

Having asked a question, selected a model system, chosen a scientific approach and obtained an answer, it must be communicated to other scientists. Peer review exists to ensure that only work of a certain quality gets into journals, but other side-effects are created. The process of peer review has been studied by psychologists on several occasions. The results are disconcerting – what partly, and perhaps largely, determines whether research is published is: (1) the perceived status of the authors; and (2) the consistency of the results with orthodox views. Consider the following studies.

D. P. Peters and Ceci (1980, 1982) postulated that professional status and institutional affiliation influenced the probability that a manuscript would be published. Their study consisted of taking articles which had already been published in mainstream journals, retyping them with fictitious names and institutions, and re-submitting them to the very journals which had initially published them. All articles had appeared in those journals within 18 to 30 months of the time they were re-submitted. All were published by individuals at prestigious institutions, and were re-submitted under low-prestige names and institutions. Out of 30 editors and referees, only 10% (three out of 30) detected the fraud. Of the remaining 10 papers, nine were rejected. Ceci and Peters suggest that scientists, like other individuals, are impressed, perhaps even at an subconscious level, by the status of others. They therefore give the benefit of the doubt to individuals presumed to have high prestige.

In another study, Mahoney (1976) did a study in which manuscripts on a controversial topic were submitted to referees. Fictitious data were reported from an experimental study with appropriate controls. In one set the data supported the current opinion. In the other set the experimental designs and the data were identical, but the labels on the data curves were reversed, leading to results which contradicted orthodoxy. In a sample of 57 individuals, the data which supported current opinion led to much more positive reviews. Interestingly enough, Mahoney had his manuscript reporting the results of this study rejected by 12 different journals. One journal which dealt primarily with physical science agreed to publish it only if Mahoney conceded to qualify it as applying only to social scientists. When he insisted on stating that it perhaps also applied to physicists and chemists, it was rejected by that journal, too.

None of the above authors suggest conscious favouratism as the mechanism – this is what makes their results of particular interest. We have already seen that there are a variety of subconscious (or, perhaps more accurately, preconscious) activities which could account for such behaviour even in individuals who believe themselves to be objective. Of course, conscious favouratism occurs, too. Invited book chapters, for example, provide a way of selecting who shall be allowed to air their views on particular topics (Table 8.2). Ceci and Peters uncovered a particularly flagrant example of deliberate favouratism in their study (Canadian Broadcasting Corporation, 1982). They had, as described above, sent a manuscript to a journal where it had just been published. The editor rejected it, but sent no referees' comments. When Ceci and Peters wrote asking for them, the editor conceded that he had personally reviewed the article and believed it had no place in the journal. They then wrote to the original author who had published the article in that very journal, and asked for copies of his reviews. The letter from the editor began, 'Dear Jim: I am happy to report that your work will appear in our journal ...'. Although many scientists find the case of such personal and conscious favouritism the most odious, it is the former examples of subconscious bias which are probably the most widespread and difficult to guard against. Possible changes to the refereeing process have been discussed elsewhere (Mahoney, 1976; Harnad, 1979; D. P. Peters and Ceci, 1980) but are not pursued here. These examples are presented as data, not as appeals to solve particular technical problems in the publication process.

It remains for sociologists to assess the influence of model systems upon referees' decisions. I predict that certain systems are far more publishable than others. Papers using birds on tropical islands, for example, may be more likely to be published than papers looking at soil invertebrates or fungi consuming leaf litter. This is a hypothesis, and I invite the test.

8.7 CONCLUSION

This is an exciting period in which to study ecology in general and competition specifically. Apart from the sheer intellectual challenge, this is a time when the world desperately needs simple, general, predictive ecological theory for conservation and sustained utilization of the biosphere. Intellectual and ethical pursuits can therefore be combined in a most satisfactory way.

Our paths to progress require a critical re-evaluation of the current situation in ecological research. Let us consider some of the possibilities for research paths in the coming decades. We can begin by specifying our goals more clearly. The mere act of specifying them, as I have tried to do above, makes the questions which we need to answer more obvious. We can ensure that our work is driven by questions, and not habitual patterns of data collection for its own sake; Table 4.4 provides guidelines for the selection of questions. In seeking to answer these questions, we can critically evaluate existing conceptual approaches to avoid the risk of persisting in blind avenues of enquiry. In Chapters 5–7, I have sketched three conceptual approaches which I believe offer potential for answering important questions more rapidly than current approaches.

In Chapter 5, I suggested planning field experiments in order to increase the generality of their conclusions. This could overcome the tendency to collect special cases, with the attendant problems of finding general principles through induction (Chapter 2). In Chapter 6, I suggest that we place more emphasis on examining interactions among many species simultaneously, and propose some questions and methods for exploring the structure of experimentally produced community matrices. This contrasts with past approaches which examine only a few populations in different environments. In Chapter 7, I argue for better-planned and better-executed comparative studies. In particular, I advocate that they be carried out using the rigorous approaches of empirical ecology and be combined with field experiments which serve as bioassays of different ecological factors.

Given the exciting opportunities that exist, what are the pitfalls and obstacles? In this chapter I explore some of the obstacles to innovation and progress in the belief that the more we are aware of such obstacles, the greater the opportunities for overcoming them are. These obstacles have largely to do with the psychology of scientists rather than the nature of ecological systems. It is obvious that progress towards competition theory (and other ecological theory) will not occur by collecting another data set consistent with established dogma, by rhetorical debates, or by doing the same thing one's supervisor did (or does). Some of the above obstacles can be consciously avoided once their existence is documented. Conscious thought, rather than tradition and habit, could guide the selection of question, system and conceptual approach. If a further recipe for progress is needed, perhaps it can come from a cookery book (Kushi, 1978):

> Do not strive to become a 'good cook' or try to change food according to your own ego. Rather think of the native qualities of each food. A cook should be like the conductor of a fine orchestra, trying to bring out all the inherent harmonies and order in the foods, their natural taste and power.

8.8 QUESTIONS FOR DISCUSSION

1. Is there presently a long-term goal for research on competition? What should it be?

2. Is there really a dichotomy between theoretical and applied questions? Discuss this in the context of the tendency of humans to erect dichotomies. What are some big applied questions that can stimulate theory?

3. Is the study of the behaviour of scientists an important part of science itself?

4. Do scientists' short-term personal goals differ from the long-term goals of their discipline?

5. Which of the potential obstacles is the greatest problem? Can you suggest simple technical methods for surmounting it? (As one example, could bias in referees be overcome by making manuscripts anonymous?) Read and discuss the suggestions of Mahoney (1976), Harnad (1979) and D. P. Peters and Ceci (1980).

6. Is 'safe' work more easily published than innovative work? Is this a testable hypothesis? Does this have consequences for the types of research done?

7. Why do audiences tend to react negatively when issues about the behaviour of scientists are raised?

8. Can we enhance creativity and falsifiability simultaneously?

9. Which important area of competition studies do you think was least well-covered in this book? Why should it have been included, and how should it have been done?

10. If you were to write an extra chapter for this book, what new topics would you cover, and how would you organize them?

References

Aarssen, L. W. (1983) Ecological combining ability and competitive combining ability in plants: towards a general evolutionary theory of coexistence in systems of competition. *The American Naturalist*, **122**, 707–31.

Aarssen, L. W. (1985) Interpretation of the evolutionary consequences of competition in plants: an experimental approach. *Oikos*, **45**, 99–100.

Aarssen, L. W. (1988) 'Pecking order' of species from pastures of different ages. *Oikos*, **51**, 3–12.

Aarssen, L. W. and Turkington, R. (1985) Competitive relations among species from pastures of different ages. *Canadian Journal of Botany*, **63**, 2319–25.

Agren, G. I. and Fagerstrom, T. (1984) Limiting dissimilarity in plants: randomness prevents exclusion of species with similar competitive abilities. *Oikos*, **43**, 369–75.

Allen, T. F. H. and Starr, T. B. (1982) *Hierarchy*, University of Chicago Press, Chicago.

Al-Mufti, M. M., Sydes, C. L., Furness, S. B., Grime, J. P. and Band, S. R. (1977) A quantitative analysis of shoot phenology and dominance in herbaceous vegetation. *Journal of Ecology*, **65**, 759–91.

Arthur, W. (1982) The evolutionary consequences of interspecific competition. *Advances in Ecological Research*, **12**, 127–87.

Arthur, W. (1987) *The Niche in Competition and Evolution*, Wiley, Chichester.

Askew, R. R. (1971) *Parasitic Insects*, Heinemann Educational, London.

Aune, B. (1970) *Rationalism, Empiricism, and Pragmatism*, Random House, New York.

Austin, M. P. (1982) Use of a relative physiological performance value in the prediction of performance in multispecies mixtures from monoculture performance. *Journal of Ecology*, **70**, 559–70.

Austin, M. P. (1986) The theoretical basis of vegetation science. *Trends in Ecology and Evolution*, **1**, 161–4.

Austin, M. P. and Austin, B. O. (1980) Behaviour of experimental plant communities along a nutrient gradient. *Journal of Ecology*, **68**, 891–918.

Baker, R. R. (1983) Insect territoriality. *Annual Review of Entomology*, **28**, 65–89.

Barrett, G. W. and Rosenberg, R. (eds) (1981) *Stress Effects on Natural Ecosystems*, Wiley, Chichester.

Bartholomew, G. A. and Heinrich, H. (1978) Endothermy in African dung beetles during flight, ball making and ball rolling. *Journal of Experimental Biology*, **73**, 65–83.

Bateson, G. (1979) *Mind and Nature*, Dutton, New York.

Begon, M. and Mortimer, M. (1981) *Population Ecology. A Unified Study of Animals and Plants*, Blackwell, Oxford.

Bender, E. A., Case, T. J. and Gilpin, M. E. (1984) Perturbation experiments in community ecology: theory and practice. *Ecology*, **65**, 1–13.

Berendse, F. B., Beltman, B., Bobbink, R., Kwant, R. and Schmitz, M. (1987a) Primary production and nutrient availability in wet heathland ecosystems. *Acta Oecologica*, **8**, 265–79.

Berendse, F. B., Oudhof, H. and Bol, J. (1987b) A comparative study on nutrient cycling in wet heathland ecosystems. I. Litter production and nutrient losses from the plant. *Oecologia (Berlin)*, **74**, 174–84.

Blum, M. S. (1981) *Chemical Defenses of Arthropods*, Academic Press, New York.

Boucher, D. H., James, S. and Keeler, K. H. (1982) The ecology of mutualism. *Annual Review of Ecology and Systematics*, **13**, 315–47.

Brown, J. H. and Davidson, D. W. (1977) Competition between seed-eating rodents and ants in desert ecosystems. *Science*, **196**, 880–2.

Brown, J. H. and Davidson, D. W. (1986) Reply to Galindo. *Ecology*, **67**, 1423–5.

Brown, J. H., Davidson, D. W., Munger, J. C. and Inouye, R. S. (1986) Experimental community ecology: the desert granivore system. In *Community Ecology* (ed. J. Diamond and T. J. Case), Harper and Row, New York, pp. 41–61.

Brown, J. H., Davidson, D. W. and Reichman, O. J. (1979) An experimental study of competition between seed-eating desert rodents and ants. *American Zoologist*, **19**, 1129–43.

Brown, J. H. and Mauer, B. A. (1986) Body size, ecological dominance and Cope's rule. *Nature*, **324**, 248–50.

Burdon, J. J. (1982) The effect of fungal pathogens on plant communities. In *The Plant Community as a working Mechanism* (ed. E. I. Newman), Blackwell, Oxford, pp. 99–112.

Buss, L. W. (1979) Habitat selection, directional growth and spatial refuges: why colonial animals have more hiding places. In *Biology and Systematics of Colonial Organisms* (eds G. Larwood and B. R. Rosen), Academic Press, London, pp. 459–97.

Buss, L. W. (1980) Competitive intransitivity and size-frequency distributions of interacting populations. *Proceedings of the National Academy of Sciences USA*, **77**, 5355–9.

Buss, L. W. (1981) Group living, competition and the evolution of cooperation in a sessile invertebrate. *Science*, **213**, 1012–4.

Buss, L. W. (1988) *The Evolution of Individuality*, Princeton University Press, Princeton, New Jersey.

Canadian Broadcasting Corporation (1982) Science and deception: Parts 1–4. *Ideas*, October 1982, CBC Transcripts, Montreal, Quebec.

Capra, F. (1975) *The Tao of Physics*, Shambala, Berkeley, California.

Capra, F. and Spretnak, C. (1984) *Green Politics*, E. P. Dutton, New York.

Chapin, F. S. III (1980) The mineral nutrition of wild plants. *Annual Review of Ecology and Systematics*, **11**, 233–60.

Chapin, F. S. III, Vitousek, P. M. and Van Cleve, K. (1986) The nature of nutrient limitation in plant communities. *The American Naturalist*, **127**, 48–58.

Chapman, R. L. (1977) *Roget's International Thesaurus.* (4th ed.) Fitzhenry and Whiteside, Toronto, Ontario.

Chesson, P. L. (1986) Environmental variation and the coexistence of species. In *Community Ecology* (ed J. Diamond and I. J. Case), Harper and Row, New York, pp. 240–56.

Chesson, P. L. and Case, T. J. (1986) Overview: Nonequilibrium community theories: chance, variability, history and coexistence. In *Community Ecology* (eds J. Diamond and T. J. Case), Harper and Row, New York, pp. 229–39.

Chomsky, N. (1987) *Pirates and Emperors*, Black Rose Books, Montreal.

Chomsky, N. and Herman, E. S. (1979) *The Washington Connection and Third World Fascism*, Black Rose Books, Montreal.

Clatworthy, J. N. and Harper, J. L. (1962) The comparative biology of closely related species living in the same area. V. Inter- and intraspecific interference within cultures of *Lemna* spp. and *Salvinia natans*. *Journal of Experimental Botany*, **13**, 307–24.

Clements, F. E. (1933) Competition in plant societies. *News Service Bulletin*, Carnegie Institution of Washington, 2 April 1933.

Clements, F. E. (1935) Experimental ecology in the public service. *Ecology*, **16**, 342–63.

Clements, F. E., Weaver, J. E. and Hanson, H. C. (1929) *Plant Competition*. Carnegie Institution of Washington, Washington, DC.

Cody, M. L. (1974) *Competition and the Structure of Bird Communities*, Princeton University Press, Princeton, New Jersey.

Cody, M. L. and Diamond, J. M. (eds) (1975) *Ecology and Evolution of Communities*, Belknap Press, Harvard University Press, Cambridge.

Coley, P. D., Bryant, J. P. and Chapin, F. S. (1985) Resource availability and plant antiherbivore defence. *Science*, **230**, 895–9.

Colinvaux, P. (1986) *Ecology*, Wiley, New York.

Collier, B. D., Cox, G. W., Johnsom, A. W. and Miller, P. C. (1973) *Dynamic Ecology*, Prentice–Hall, Englewood Cliffs, New Jersey.

Colwell, R. K. and Fuentes, E. R. (1975) Experimental studies of the niche. *The Annual Review of Ecology and Systematics*, **6**, 281–309.

Connell, J. H. (1961) The influence of interspecific competition and other factors on the distribution of the barnacle *Chthamalus stellatus*. *Ecology*, **42**, 710–23.

Connell, J. H. (1972) Community interactions on marine rocky intertidal shores. *Annual Review of Ecology and Systematics*, **3**, 169–92.

Connell, J. H. (1975) Some mechanisms producing structure in natural communities: a model and evidence from field experiments. In *Ecology and Evolution of Communities* (eds M. L. Cody and J. M. Diamond), Belknap Press, Harvard University Press, Cambridge, pp. 460–90.

Connell, J. H. (1978) Diversity in tropical rain forests and coral reefs. *Science*, **199**, 1302–10.

Connell, J. H. (1980) Diversity and the coevolution of competitors, or the ghost of competition past. *Oikos*, **35**, 131–8.

Connell, J. H. (1983) On the prevalence and relative importance of interspecific competition: evidence from field experiments. *The American Naturalist*, **122**, 661–96.

Connell, J. H. (1989) Apparent vs. 'real' competition in plants. In *Perspectives on Plant Competition* (eds J. Grace and D. Tilman), Academic Press, New York.

Connor, E. F. and Simberloff, D. (1979) The assembly of species communities: chance or competition? *Ecology*, **60**, 1132–40.

Crick, J. C. and Grime, J. P. (1987). Morphological plasticity and mineral nutrient capture in two herbaceous species of contrasted ecology. *New Phytologist*, **107**, 403–14.

Crowson, R. A. (1981) *The Biology of the Coleoptera*, Academic Press, London.

Cummins, K. W. (1973) Trophic relationships of aquatic insects. *Annual Review of Entomology*, **18**, 183–206.

Cummins, K. W. and Klug, M. J. (1979) Feeding ecology of stream invertebrates. *Annual Review of Ecology and Systematics*, **10**, 147–72.

Currie, D. J. and Paquin, V. (1987) Large-scale biogeographical patterns of species richness of trees. *Nature*, **329**, 326–7.

Dale, M. R. T. (1984) The contiguity of upslope and downslope boundaries of species in a zoned community. *Oikos*, **42**, 92–6.

Damuth, J. (1981) Population density and body size in mammals. *Nature*, **290**, 699–700.

Darwin, C. (1859) *On the Origin of Species by Means of Natural Selection*, Reprinted 1958 by New American Library, New York.

Dawkins, R. (1976) *The Selfish Gene*, Oxford University Press, Oxford.

Day, R. T., Keddy, P. A., McNeill, J. and Carleton, T. (1988) Fertility and disturbance gradients: a summary model for riverine marsh vegetation. *Ecology*, **69**, 1044–54.

Dayton, P. K. (1975) Experimental evaluation of ecological dominance in a rocky intertidal algal community. *Ecological Monographs*, **45**, 137–59.

Dayton, P. K. (1979) Ecology: a science and a religion. In *Ecological Processess in Coastal and Marine Systems* (ed. R. J. Livingston), Plenum Press, New York, pp. 3–18.

de Wit, C. T. (1960) On competition. *Verslagen van Landbouwkundige Onderzoekingen*, **66**, 1–82.

del Moral, R. (1983) Competition as a control mechanism in subalpine meadows. *American Journal of Botany*, **70**, 232–45.

Diamond, J. M. (1975) Assembly of species communities. In *Ecology and Evolution of Communities* (eds M. L. Cody and J. M. Diamond), Belknap Press, Harvard University Press, Cambridge, pp. 342–444.

Diamond, J. M. (1983) Laboratory, field and natural experiments. *Nature*, **304**, 586–7.

Diamond, J. M. (1986) Overview: Laboratory experiments, field experiments and natural experiments. In *Community Ecology* (eds J. Diamond and T. J. Case), Harper and Row, New York, pp. 3–22.

Diamond, J. M. and Case, T. J. (eds) (1986) *Community Ecology*, Harper and Row, New York.

Diamond, J. M. and Gilpin, M. (1982) Examination of the 'null' model of Connor and Simberloff for species co-occurrences on islands. *Oecologia*, **52**, 64–74.

Digby, P. G. N. and Kempton, R. A. (1987) *Multivariate Analysis of Ecological Communities*, Chapman and Hall, London.

du Rietz, G. E. (1931) Life-forms of terrestrial flowering plants. *Acta Phytogeographica Suecica*, *III*, Almqvist and Wiksells, Uppsala.

Dye, T. R. and Zeigler, H. (1987) *The Irony of Democracy*, 7th edn, Brooks/Cole, Monterey, California.

Ehrlich, P. R. and Birch, L. C. (1967) The 'balance of nature' and 'population control'. *The American Naturalist*, **101**, 97–107.

Ehrlich, P. R. and Ehrlich, A. H. (1981) *Extinction*, Random House, New York.

Emmons, L. H. (1980) Ecology and resource partitioning among nine species of African rain forest squirrels. *Ecological Monographs*, **50**, 31–54.

Fagerstrom, T. (1987) On theory, data and mathematics in ecology. *Oikos*, **50**, 258–61.

Feeny, P. (1976) Plant apparency and chemical defense. *Recent Advances in Phytochemistry*, **10**, 1–40.

Ferson, S., Downey, P., Klerks, P., Weissburg, M., Kroot, I., Stewart, S., Jacquez, G., Ssemakula, J., Malenky, R. and Anderson, K. (1986) Competing reviews, or why do Connell and Schoener disagree? *The American Naturalist*, **127**, 571–6.

Firbank, L. G. and Watkinson, A. R. (1985) On the analysis of competition within two-species mixtures of plants. *Journal of Applied Ecology*, **22**, 503–17.

Fisher, R. C. (1961) A study in insect multiparasitism. II. The mechanism and control of competition for possession of the host. *Journal of Experimental Biology*, **38**, 605–28.

Fitter, A. H. and Hay, R. K. M. (1981) *Environmental Physiology of Plants*, Academic Press, London.

Fitzpatrick, S. M. and Wellington, W. G. (1983) Insect territoriality. *Canadian Journal of Zoology*, **61**, 471–86.

Fonteyn, P. J. and Mahall, B. E. (1981) An experimental analysis of structure in a desert plant community. *J. Ecology*, **69**, 883–96.

Fowler, N. (1981) Competition and coexistence in a North Carolina grassland. II. The effects of the experimental removal of species. *The Journal of Ecology*, **69**, 843–5.

Fowler, N. (1986) The role of competition in plant communities in arid and semiarid regions. *Annual Review of Ecology and Systematics*, **17**, 89–110.

Friedenberg, E. Z. (1976) *The Disposal of Liberty and Other Industrial Wastes*, Doubleday, Garden City.

Galindo, C. (1986) Do desert rodent populations increase when ants are removed? *Ecology*, **67**, 1422–3.

Gauch, H. G. (1982) *Multivariate Analysis in Community Ecology*, Cambridge University Press, Cambridge.

Gaudet, C. L. and Keddy, P. A. (1988) Predicting competitive ability from plant traits: a comparative approach. *Nature*, **334**, 242–3.

Gause, G. F. (1932) Experimental studies on the struggle for existence. I. Mixed population of two species of yeast. *Journal of Experimental Biology*, **9**, 389–402.

Gause, G. F. (1934) *The Struggle for Existence*, Hafner, New York.

Giller, P. S. (1984) *Community Structure and the Niche*, Chapman and Hall, London.

Gilpin, M. E. and Diamond, J. M. (1982) Factors contributing to the non-randomness in species co-occurrences on islands. *Oecologia*, **52**, 75–84.

Gilpin, M. E., Carpenter, M. P. and Pomerantz, M. J. (1986) The assembly of a laboratory community: multispecies competition in *Drosophila*. In *Community Ecology* (eds J. Diamond and T. J. Case), Harper and Row, New York, pp. 23–40.

Goldberg, D. E. (1987) Neighbourhood competition in an old-field plant community. *Ecology*, **68**, 1211–23.

Goldberg, D. E. and Fleetwood, L. (1987) Competitive effect and response in four annual plants. *Journal of Ecology*, **75**, 1131–43.

Goldsmith, F. B. (1978) Interaction (competition) studies as a step towards the synthesis of seacliff vegetation. *Journal of Ecology*, **66**, 921–31.

Goldstein, J. (1983) *The Experience of Insight*, Shambala, Boston.

Gorham, E. (1979) Shoot height, weight and standing crop in relation to density of nonspecific plant stands. *Nature*, **274**, 148–50.

Gould, S. J. (1981) *The Mismeasure of Man*, W. W. Norton, New York.

Gould, S. J. (1983) *Hen's Teeth and Horse's Toes*, W. W. Norton, New York.

Grace, J. B. (1987) The impact of preemption on the zonation of two *Typha* species along lakeshores. *Ecological Monographs*, **57**, 283–303.

Grace, J. B. (1989) On the relationship between plant traits and competitive ability. In *Perspectives on Plant Competition* (eds J. Grace and D. Tilman), Academic Press, New York.

Grace, J. B. and Wetzel, R. G. (1981) Habitat partitioning and competitive displacement in cattails (*Typha*): experimental field studies. *The American Naturalist*, **118,** 463–74.

Grant, P. R. and Abbott, I. (1980) Interspecific competition, island biogeography and null hypotheses. *Evolution*, **34,** 332–41.

Green, R. H. (1980) Multivariate approaches in ecology: the assessment of ecologic similarity. *Annual Review of Ecology and Systematics*, **11,** 1–14.

Greenslade, P. J. M. (1983) Adversity selection and the habitat templet. *The American Naturalist*, **122,** 352–65.

Grime, J. P. (1973) Competitive exclusion in herbaceous vegetation. *Nature*, **242,** 344–7.

Grime, J. P. (1974) Vegetation classification by reference to strategies. *Nature*, **250,** 26–31.

Grime, J. P. (1977) Evidence for the existence of three primary strategies in plants and its relevance to ecological and evolutionary theory. *The American Naturalist*, **111,** 1169–94.

Grime, J. P. (1979) *Plant Strategies and Vegetation Processes*, Wiley, Chichester.

Grime, J. P. and Hunt, R. (1975) Relative growth rate: its range and adaptive significance in a local flora. *Journal of Ecology*, **63,** 393–422.

Grime, J. P., Mason, G., Curtis, A. V., Rodman, J., Band, S. R., Mowforth, M. A. G., Neal, A. M. and Shaw, S. (1981) A comparative study of germination characteristics in a local flora. *Journal of Ecology*, **69,** 1017–59.

Grubb, P. J. (1977) The maintenance of species richness in plant communities: the importance of the regeneration niche. *Biological Reviews*, **52,** 107–45.

Grubb, P. J. (1985) Plant populations and vegetation in relation to habitat, disturbance and competition: problems of generalization. In *The Population Structure of Vegetation* (ed. J. White), Junk, Dordrecht, pp. 595–621.

Gurevitch, J. (1986) Competition and the local distribution of the grass *Stipa neomexicana*. *Ecology*, **67,** 46–57.

Haefner, J. W. (1978) Ecosystem assembly grammars: generative capacity and empirical adequacy. *Journal of Theoretical Biology*, **73,** 293–318.

Haefner, J. W. (1981) Avian community assembly rules: the foliage-gleaning guild. *Oecologia (Berlin)*, **50,** 131–42.

Hairston, N. G., Smith, F. E. and Slobodkin, L. B. (1960) Community structure, population control, and competition. *The American Naturalist*, **94,** 421–5.

Haldane, J. B. S. (1985) *On Being the Right Size* (ed. J. Maynard Smith), Oxford University Press, Oxford.

Hardin, G. (1960) The competitive exclusion principle. *Science*, **131,** 1292–7.

Harnad, S. (1979) Creative disagreement. *The Sciences*, (September), 18–20.

Harper, J. L. (1960) The evolution and ecology of closely related species living in the same area. *Evolution*, **15,** 209–27.

Harper, J. L. (1961) Approaches to the study of plant competition. *Symposia of the Society for Experimental Biology*, **15**, 1–39.

Harper, J. L. (1965) The nature and consequence of interference amongst plants. *Genetics Today*, **2**, 465–82.

Harper, J. L. (1977) *Population Biology of Plants*, Academic Press, London.

Harper, J. L. (1982) After description. In *The Plant Community as a Working Mechanism* (ed. E. I. Newman), Blackwell, Oxford, pp. 11–25.

Harper, J. L. and Chancellor, A. P. (1959) The comparative biology of closely related species living in the same area. IV. *Rumex*: interference between individuals in populations of one and two species. *Journal of Ecology*, **47**, 679–95.

Harper, J. L. and Clatworthy, J. N. (1963) The comparative biology of closely related species. VI. Analysis of the growth of *Trifolium repens* and *T. frageriferum* in pure and mixed populations. *Journal of Experimental Botany*, **4**, 172–90.

Harper, J. L. and McNaughton, J. H. (1962) The comparative biology of closely related species living in the same area. VII. Interference between individuals in pure and mixed populations of *Papaver* species. *New Phytologist*, **61**, 175–88.

Hayward, J. M. (1987) *Shifting Worlds, Changing Minds*, New Science Library, Shambala, Boulder, Colorado.

Heinrich, B. and Bartholomew, G. A. (1979) Roles of endothermy and size in inter- and intraspecific competition for elephant dung in an African dung beetle, *Scarabaeus laevistriatus*. *Physiological Zoology*, **52**, 484–96.

Herbold, B. and Moyle, P. B. (1986) Introduced species and vacant niches. *The American Naturalist*, **128**, 751–60.

Hicks, C. R. (1964) *Fundamental Concepts in the Design of Experiments*, Holt, Rinehart and Winston, New York.

Hofstadter, D. R. and Dennett, D. C. (1981) *The Mind's I*, Basic Books, New York.

Holling, C. S. (ed.) (1978) *Adaptive Environmental Assessment and Management*, Wiley, Chichester.

Holt, R. D. (1977) Predation, apparent competition, and the structure of prey communities. *Theoretical Population Biology*, **12**, 197–229.

Hurlbert, S. H. (1984) Pseudoreplication and the design of ecological field experiments. *Ecological Monographs*, **54**, 187–211.

Huston, M. (1979) A general hypothesis of species diversity. *The American Naturalist*, **113**, 81–101.

Huston, M. (1985) Patterns of species diversity on coral reefs. *Annual Review of Ecology and Systematics*, **16**, 149–77.

Hutchins, R. M. (1963) (reprinted 1987 as Science, Scientists and Politics) *The Centre Magazine*, **20** (6), 29–32. (Center for the Study of Democratic Institutions, University of California).

Hutchinson, G. E. (1959) Homage to Santa Rosalia or why are there so many kinds of animals? *The American Naturalist*, **93**, 145–9.

Hutchinson, G. E. (1978) *An Introduction to Population Ecology*, Yale University Press, New Haven.

International Union for the Conservation of Nature and Natural Resources (1980) *World Conservation Strategy*, IUCN, Gland, Switzerland.

Jackson, J. B. C. (1981) Interspecific competition and species distributions: the ghosts of theories and data past. *American Zoologist*, **21**, 889–901.

Janzen, D. H. (1975) Interactions of seeds and their insect predators/parasitoids in a tropical deciduous forest. In *Evolutionary Strategies of Parasitic Insects and Mites* (ed. P. W. Price), Plenum Press, New York. pp. 154–86.

Jung, C. G. (1964) *Man and His Symbols*, Doubleday, Garden City.

Karr, J. R. (1983) Commentary to Weins, J. A. Avian community ecology: an iconoclastic view. In *Perspectives in Ornithology* (eds A. H. Brush and G. A. Clark), Cambridge University Press, Cambridge, pp. 403–10.

Kaufmann, J. H. (1983) On the definitions and functions of dominance and territoriality. *Biological Reviews*, **58**, 1–20.

Keddy, P. A. (1981) Experimental demography of the sand dune annual, *Cakile edentula*, growing along an environmental gradient in Nova Scotia. *Journal of Ecology*, **69**, 615–30.

Keddy, P. A. (1982) Population ecology on an environmental gradient: *Cakile edentula* on a sand dune. *Oecologia (Berlin)*, **52**, 348–55.

Keddy, P. A. (1983) Shoreline vegetation in Axe Lake, Ontario: effects of exposure on zonation patterns. *Ecology*, **64**, 331–44.

Keddy, P. A. (1987) Beyond reductionism and scholasticism in plant community ecology. *Vegetatio*, **69**, 209–11.

Keddy, P. A. (1989a) Competitive hierarchies and centrifugal organization in plant communities. In *Perspectives on Plant Competition* (eds J. Grace and D. Tilman), Academic Press, New York.

Keddy, P. A. (1989b) Effects of competition from shrubs on herbaceous wetland plants: a four year field experiment. *Canadian Journal of Botany*, 1989, **67**.

Keddy, P. A. and Shipley, B. (1989) Competitive hierarchies in plant communities. *Oikos*, **49**.

Kershaw, K. A. (1973) *Quantitative and Dynamic Plant Ecology*, 2nd edn, Edward Arnold, London.

Khailov, K. M. (1986) Thematic structure of publications on aquatic ecology in 1984. *Ekologiya*, **4**, 71–7 (transl. from Russian).

Kirkpatrick, J. B. and Hutchinson, C. F. (1977) The community composition of Californian coastal sage scrub. *Vegetatio*, **35**, 21–33.

Klare, M. T. and Arnson, C. (1981) *Supplying Repression*, Institute for Policy Studies, Washington, DC.

Krebs, C. J. (1978) *Ecology: the Experimental Analysis of Distribution and Abundance*, 2nd edn, Harper and Row, New York.

Kuhn, T. S. (1970) *The Structure of Scientific Revolutions*, The University of Chicago Press, Chicago.

Kushi, A. T. (1978) *How to Cook with Miso*, Japan Publications, Tokyo.

Lack, D. (1966) *Population Studies of Birds*, Clarendon Press, Oxford.

Lappe, F. M. (1971) *Diet for a Small Planet*, Ballantine Books, New York.

Lappe, F. M. and Collins, J. (1982) *Food First*, Sphere Books, London.

Lawrey, J. D. (1981) Evidence for competitive release in simplified saxicolous lichen communities. *American Journal of Botany*, **68,** 1066–73.

Lawton, J. H. (1984) Non-competitive populations, non-convergent communities and vacant niches: the herbivores of bracken. In *Ecological Communities: Conceptual Issues and the Evidence* (eds D. R. Strong, D. Simberloff, L. G. Abele and A. B. Thistle), Princeton University Press, Princeton, New Jersey, pp. 67–100.

Lawton, J. H. and Hassell, M. P. (1981) Asymmetrical competition in insects. *Nature*, **289,** 793–5.

Leary, R. A. (1985) A framework for assessing and rewarding a scientist's research productivity. *Scientometrics* **7,** 29–38.

Lederer, R. J. (1984) *Ecology and Field Biology*, Benjamin/Cummings, Menlo Park, California.

Legendre, L. and Legendre, P. (1983) *Numerical Ecology*, Elsevier, Amsterdam.

Levins, R. (1968) *Evolution in Changing Environments*, Princeton University Press, Princeton, New Jersey.

Levins, R. (1975) Evolution in communities near equilibrium. In *Ecology and Evolution of Communities* (eds M. L. Cody and J. M. Diamond), Belknap Press, Harvard University Press, Cambridge, Massachusetts, pp. 16–50.

Levins, R. and Lewontin, R. (1982) Dialectics and reductionism. In *Conceptual Issues in Ecology* (ed. E. Saarinen), D. Reidel, Dordrecht, pp. 107–138.

Levitt, J. (1980) *Responses of Plants to Environmental Stresses*, 2nd edn, Academic Press, New York.

Lewontin, R. C. (1974) *The Genetic Basis of Evolutionary Change*, Columbia University Press, New York.

Loehle, C. (1987) Hypothesis testing in ecology: Psychological aspects and the importance of theory maturation. *The Quarterly Review of Biology*, **62,** 397–409.

Longstaff, B. C. (1976) The dynamics of collembolan populations: competitive relationships in an experimental system. *Canadian Journal of Zoology*, **54,** 948–62.

Lotka, A. J. (1932) The growth of mixed populations: Two species competing for a common food supply. *Journal of the Washington Academy of Sciences*, **22,** 461–9.

Lowe, M. (1981) Cooperation and competition in science. *International Journal of Women's Studies*, **4**, 362–8.

Lubchenco, J. (1980) Algal zonation in the New England rocky intertidal community: an experimental analysis. *Ecology*, **61**, 333–44.

MacArthur, R. H. (1972) *Geographical Ecology*, Harper and Row, New York.

Maccoby, E. E. and Jacklin, C. N. (1974) *The Psychology of Sex Differences*, Stanford University Press, Stanford.

McGilchrist, C. A. and Trenbath, B. R. (1971) A revised analysis of plant competition experiments. *Biometrics*, **27**, 659–71.

McIntosh, R. P. (1985) *The Background of Ecology*, Cambridge University Press, Cambridge.

McNaughton, S. J. and Wolf, L. L. (1970) Dominance and the niche in ecological systems. *Science*, **167**, 131–9.

McNaughton, S. J. and Wolf, L. L. (1979) *General Ecology*, 2nd edn, Holt, Rinehart and Winston, New York.

Magee, B. (1973) *Popper*, Fontana Press, London.

Mahoney, M. J. (1976) *Scientist as Subject: The Psychological Imperative*, Ballinger, Cambridge, Mass.

Mann, K. H. (1985) Invertebrate behaviour and the structure of marine benthic communities. In *Behavioural Ecology* (eds R. M. Sibley and R. H. Smith) Blackwell, Oxford, pp. 227–46.

Margulis, L. (1970) *Origin of Eucaryotic Cells*, Yale University Press, New Haven, Connecticut.

Matthews, G. J. and Morrow, R. (1985) *Canada and the World, An Atlas Resource*, Prentice–Hall Canada, Scarborough.

May, R. M. (1974) *Stability and Complexity in Model Ecosystems*, 2nd edn, Princeton University Press, Princeton, New Jersey.

May, R. M. (1981) Patterns in multi-species communities. In *Theoretical Ecology*, 2nd edn (ed. R. M. May), Blackwell, Oxford, pp. 197–227.

May, R. M. (1986) The search for patterns in the balance of nature: advances and retreats. *Ecology*, **67**, 1115–26.

Melrose, D. (1985) *Nicaragua: the Threat of a Good Example?* Oxfam, UK.

Miller, R. S. (1967) Pattern and process in competition. *Advances in Ecological Research*, **4**, 1–74.

Miller, T. E. (1982) Community diversity and interactions between the size and frequency of disturbance. *The American Naturalist*, **120**, 533–6.

Milne, A. (1961) Definition of competition among animals. *Symposia of the Society for Experimental Biology*, **15**, 40–61.

Mitchley, J. (1988) Control of relative abundance of perennials in chalk grassland in southern England. II. Vertical canopy structure. *Journal of Ecology*, **76**, 341–50.

Mitchley, J. and Grubb, P. J. (1986) Control of relative abundance of perennials in chalk grassland in southern England. I. Constancy of rank order and results of pot- and field-experiments on the role of interference. *Journal of Ecology*, **74,** 1139–66.

Mooney, H. A. and Godron, M. (eds) (1983) *Disturbance and Ecosystems*, Springer-Verlag, Berlin.

Moore, D. R. J., Keddy, P. A., Gaudet, C. L. and Wisheu, I.C. (1989) Conservation of wetlands: do infertile wetlands deserve a higher priority? *Biological Conservation*.

Morowitz, H. J. (1968) *Energy Flow in Biology*, Academic Press, New York.

Moulton, M. P. and Pimm, S. L. (1986) The extent of competition in shaping an introduced avifauna. In *Community Ecology* (eds J. Diamond and T. J. Case), Harper and Row, New York, pp. 80–97.

Mueller-Dombois, D. and Ellenberg, H. (1974) *Aims and Methods of Vegetation Ecology*, Wiley, New York.

Murdoch, W. W. (1966) Community structure, population control, and competition – a critique. *The American Naturalist*, **100,** 219–26.

Myers, N. (1985) *The Gaia Atlas of Planet Management*, Pan Books, London.

Newman, E. I. (1973) Competition and diversity in herbaceous vegetation. *Nature*, **244,** 310.

Odum, E. P. (1983) *Basic Ecology*, Saunders College Publishing, Philadelphia.

Orgel, L. E. (1973) *The Origins of Life: Molecules and Natural Selection*, Wiley, New York.

Orloci, L. (1978) *Multivariate Analysis in Vegetation Research*, 2nd edn, Junk, The Hague.

Orwell, G. (1945) *Animal Farm*, Secker and Warburg, London.

Park, T. (1948) Experimental studies of interspecies competition. I. Competition between populations of the flour beetles, *Tribolium confusum* Duval and *Tribolium castaneum* Herbst. *Ecological Monographs*, **18,** 265–307.

Park, T. (1954) Experimental studies of interspecies competition. II. Temperature, humidity, and competition in two species of *Tribolium*. *Physiological Zoology*, **27,** 177–238.

Parker, G. A. (1984) Sperm competition and the evolution of animal mating strategies. In *Sperm Competition and the Evolution of Animal Mating Systems* (ed. R. L. Smith), Academic Press, Orlando.

Parker, M. A. and Root, R. B. (1981) Insect herbivores limit habitat distribution of a native composite *Machaeranthera canescens*. *Ecology*, **62,** 1390–2.

Persson, L. (1985) Asymmetrical competition: are larger animals competitively superior? *The American Naturalist*, **126,** 261–6.

Peters, D. P. and Ceci, S. J. (1980) A manuscript masquerade. *The Sciences*, **35,** (September), 16–9.

Peters, D. P. and Ceci, S. J. (1982) Resubmitting previously published articles: a study of the journal review process in psychology. *Behavioral and Brain Sciences*, **5,** 187–95.

Peters, R. H. (1980a) Useful concepts for predictive ecology. *Synthese*, **43,** 257–69.

Peters, R. H. (1980b) From natural history to ecology. *Perspectives in Biology and Medicine*, **23**, 191–203.

Pianka, E. R. (1973) The structure of lizard communities. *Annual Review of Ecology and Systematics*, **4**, 53–74.

Pianka, E. R. (1981) Competition and niche theory. In *Theoretical Ecology*, 2nd edn (ed. R. M. May), Blackwell, Oxford, pp. 167–96.

Pianka, E. R. (1983) *Evolutionary Ecology*, 3rd edn, Harper and Row, New York.

Pickett, S. T. A. and White, P. S. (1985) *The Ecology of Natural Disturbance and Patch Dynamics*, Academic Press, Orlando, Florida.

Picman, J. (1980) Impact of marsh wrens on reproductive strategy of red-winged blackbirds. *Canadian Journal of Zoology*, **58**, 337–50.

Picman, J. (1984) Experimental study on the role of intra- and inter-specific behaviour in marsh wrens. *Canadian Journal of Zoology*, **62**, 2353–6.

Pielou, E. C. (1972) On kinds of models. *Science*, **177**, 981–2.

Pielou, E. C. (1977) *Mathematical Ecology*, Wiley, New York.

Pielou, E. C. (1979) On A. J. Underwood's model for a random pattern. *Oecologia*, **44**, 143–4.

Pielou, E. C. (1984) *The Interpretation of Ecological Data*, Wiley, New York.

Pimm, S. L. (1978) An experimental approach to the effects of predictability on community structure. *American Zoologist*, **18**, 797–808.

Pimm, S. L. (1982) *Food Webs*, Chapman and Hall, London.

Pimm, S. L. and Pimm, J. L. (1982) Resource use, competition, and resource availability in Hawaiian Honeycreepers. *Ecology*, **63**, 1468–80.

Pimm, S. L. and Rosenzweig, M. L. (1981) Competitors and habitat use. *Oikos*, **37**, 1–6.

Pirsig, R. M. (1974) *Zen and the Art of Motorcycle Maintenance*, Morrow, New York.

Platt, J. R. (1964) Strong inference. *Science*, **146**, 347–53.

Platt, W. J. and Weis, I. M. (1985) An experimental study of competition among fugitive prairie plants. *Ecology*, **66**, 708–20.

Popper, K. (1959) *The Logic of Scientific Discovery*, Basic Books, New York.

Price, P. W. (1984a) Alternative paradigms in community ecology. In *A New Ecology: Novel Approaches to Interactive Systems* (eds P. W. Price, C. N. Slobodchikoff and W. S. A. Gaud), Wiley, New York, pp. 354–83.

Price P. W. (1984b) Communities of specialists: vacant niches in ecological and evolutionary time. In *Ecological Communities: Conceptual Issues and the Evidence* (eds D. R. Strong, D. Simberloff, L. G. Abele and A. B. Thistle), Princeton University Press, Princeton, New Jersey, pp. 510–23.

Prigogine, I. and Stengers, I. (1984) *Order Out of Chaos*, New Science Library, Shambala, Boulder, Colorado.

Putwain, P. D. and Harper, J. L. (1970) Studies in the dynamics of plant populations. III. The influence of associated species on populations of *Rumex acetosa* L. and *R. acetosella* L. in grassland. *Journal of Ecology*, **58**, 251–64.

Pyke, G. H. (1984) Optimal foraging theory: a critical review. *Annual Review of Ecology and Systematics*, **15**, 523–75.

Quinn, J. F. and Dunham, A. E. (1983) On hypothesis testing in ecology and evolution. *The American Naturalist*, **122**, 602–17.

Rabinowitz, D. (1978) Early growth of mangrove seedlings in Panama, and an hypothesis concerning the relationship of dispersal and zonation. *Journal of Biogeography*, **5**, 113–33.

Rhodes, A. (1976), *Propaganda* (ed. V. Margolin), Chelsea House, New York.

Ricklefs, R. E. (1976) *Ecology*, 2nd edn, Chiron Press, New York.

Ricklefs, R. E. (1983) *The Economy of Nature*, Chiron Press, New York.

Rigler, F. H. (1982) Recognition of the possible: an advantage of empiricism in ecology. *Canadian Journal of Fisheries and Aquatic Sciences*, **39**, 1323–31.

Root, R. (1967) The niche exploitation pattern of the blue-grey gnatcatcher. *Ecological Monographs*, **37**, 317–50.

Rorison, I. H., Grime, J. P., Hunt, R., Hendry, G. A. F. and Lewis, D. H. (1987) Frontiers of comparative plant ecology. *New Phytologist*, **106** (supplement), 1–317.

Rosenzweig, M. L. (1979) Three probable evolutionary causes for habitat selection. In *Contemporary Quantitative Ecology and Related Ecometrics* (eds G. P. Patil and M. Rosenzweig), International Co-operative Publishing House, Fairland, pp. 49–60.

Rosenzweig, M. L. (1981) A theory of habitat selection. *Ecology*, **62**, 327–5.

Rosenzweig, M. L. and Abramsky, Z. (1986) Centrifugal community organization. *Oikos* **46**, 339–48.

Rothschild, J. (ed.) (1983) *Machina Ex Dea*, Pergamon Press, New York.

Roughgarden, J. (1979) *Theory of Population Genetics and Evoloutionary Ecology: an Introduction*, MacMillan, New York.

Roughgarden, J. (1983) Coevolution between competitors. In *Coevolution* (eds D. J. Futuyma and M. Slatkin), Sinauer, Sunderland, pp. 383–403.

Saarinen, E. (ed.) (1982) *Conceptual Issues in Ecology*, D. Reidel, Dordrecht.

Salt, G. (1961) Competition among insect parasitoids. In *Mechanisms in Biological Competition. Symposia of the Society for Experimental Biology*, **15**, 98–119.

Salt, G. (1983), Roles: their limits and responsibilities in ecological and evolutionary research. *The American Naturalist*, **122**, 697–705.

Sanders, H. L. (1968) Marine benthic diversity: a comparative study. *The American Naturalist*, **102**, 243–82.

Schoener, T. W. (1974) Resource partitioning in ecological communities. *Science*, **185**, 27–39.

Schoener, T. W. (1983) Field experiments on interspecific competition. *The American Naturalist*, **122**, 240–85.

Schoener, T. W. (1986) Overview: kinds of ecological communities – ecology becomes pluralistic. In *Community Ecology* (eds J. Diamond and T. J. Case), Harper and Row, New York, pp. 467–79.

Sell, D. W., Carney, H. J. and Fahnenstiel, G. L. (1984) Inferring competition between natural phytoplankton populations: the Lake Michigan examples reexamined. *Ecology*, **65**, 325–8.

Severinghaus, W. D. (1981) Guild theory development as a mechanism for assessing environmental impact. *Environmental Management*, **5**, 187–90.

Sharitz, R. R. and McCormick J. F. (1973) Population dynamics of two competing annual plant species. *Ecology*, **54**, 723–40.

Shipley, B. (1987) Pattern and mechanism in the emergent macrophyte communities along the Ottawa River (Canada). PhD thesis, University of Ottawa.

Shipley, B. and Keddy, P. A. (1987) The individualistic and community-unit concepts as falsifiable hypotheses. *Vegetatio*, **69**, 47–55.

Siefert, R. P. and Siefert, F. H. (1976) A community matrix analysis of *Heliconia* insect communities. *The American Naturalist*, **110**, 461–83.

Siegel, S. (1956) *Nonparametric Statistics for the Behavioral Sciences*, McGraw-Hill, New York.

Silander, J. A. and Antonovics, J. (1982) Analysis of interspecific interactions in a coastal plant community – a perturbation approach. *Nature*, **298**, 557–60.

Silvertown, J. (1980) The dynamics of a grassland ecosystem: botanical equilibrium in the park grass experiment. *Journal of Applied Ecology*, **17**, 491–504.

Silvertown, J. (1987) *Introduction to plant population ecology*, 2nd edn, Longman, London.

Simberloff, D. (1983a) Competition theory, hypothesis testing, and other community ecological buzzwords. *The American Naturalist*, **122**, 626–35.

Simberloff, D. (1983b) Sizes of coexisting species. In *Coevolution* (eds D. J. Futuyma and M. Slatkin), Sinauer, Sunderland, pp. 404–30.

Simberloff, D. (1984) The great god of competition. *The Sciences*, **24**, 17–22.

Sindermann, C. J. (1982) *Winning the Games Scientists Play*, Plenum Press, New York.

Slobodkin, L. B., Smith, F. E. and Hairston, N. G. (1967) Regulation in terrestrial ecosystems, and the implied balance of nature. *The American Naturalist*, **101**, 109–24.

Smith, R. L. (1984) Human sperm competition. In *Sperm Competition and the Evolution of Animal Mating Systems* (ed. R. L. Smith), Academic Press, Orlando, Florida.

Smith, R. L. (1986) *Elements of Ecology*, 2nd edn, Harper and Row, New York.

Snow, A. A. and Vince, S. W. (1984) Plant zonation in an Alaskan salt marsh. II. An experimental study of the role of edaphic conditions. *Journal of Ecology*, **72**, 669–84.

Solbrig, O. T., Jain, S., Johnson, G. G. and Raven, P. H. (1979) *Topics in Plant Population Biology*, Columbia University Press, New York.

Sousa, W. P. (1984) The role of disturbance in natural communities. *Annual Review of Ecology and Systematics*, **15**, 353–91.

Southwood, T. R. E. (1977) Habitat, the templet for ecological strategies? *Journal of Animal Ecology*, **46**, 337–65.

Southwood, T. R. E. (1985) Interactions of plants and animals: patterns and processes. *Oikos*, **44**, 5–11.

Southwood, T. R. E. (1987) Habitat and insect biology. *Bulletin of the Entomological Society of America*, 211–14.

Southwood, T. R. E. (1988) Tactics, strategies and templets. *Oikos*, **52**, 3–18.

Southwood, T. R. E., Brown, V. K. and Reader, P. M. (1986) Leaf palatability, life expectancy and herbivore damage. *Oecologia (Berlin)*, **70**, 544–8.

Springett, B. P. (1968) Aspects of the relationship between burying beetles, *Necrophorus* spp. and the mite, *Poecilochirus necrophori* Vitz. *Journal of Animal Ecology*, **37**, 417–24.

Starfield, A. M. and Bleloch, A. L. (1986) *Building Models for Conservation and Wildlife Management*, Macmillan, New York.

Stearns, S. C. (1976) Life history tactics: a review of the ideas. *Quarterly Review of Biology*, **51**, 3–47.

Stearns, S. C. (1982) The emergence of evolutionary and community ecology as experimental sciences. *Perspectives in Biology and Medecine*, **25**, 621–48.

Strong, D. R. (1982a) Potential interspecific competition and host specificity: hispine beetles on *Heliconia*. *Ecological Entomology*, **7**, 217–20.

Strong, D. R. (1982b) Harmonious coexistence of hispine beetles on *Heliconia* in experimental and natural communities. *Ecology*, **63**, 1039–49.

Strong, D. R., Lawton, J. H. and Southwood, R. (1984) *Insects on Plants*, Harvard University Press, Cambridge, Massachusetts.

Strong, D. R., Simberloff, D., Abele, L. G. and Thistle, A. B. (eds) (1985) *Ecological Communities: Conceptual Issues and the Evidence*, Princeton University Press, Princeton, New Jersey.

Tansley, A. G. (1914) Presidential Address. *Journal of Ecology*, **2**, 194–203.

Terborgh, J. (1971) Distribution on environmental gradients: theory and a preliminary interpretation of distributional patterns in the avifauna of the Cordillera Vilcabamba, Peru. *Ecology*, **52**, 23–40.

Thompson, K. (1987) The resource ratio hypothesis and the meaning of competition. *Functional Ecology*, **1**, 297–303.

Thompson, K. and Grime, J. P. (1988) Competition reconsidered – a reply to Tilman. *Functional Ecology*, **2**, 114–16.

Tilman, D. (1982) *Resource Competition and Community Structure*, Princeton University Press, Princeton, New Jersey.

Tilman, D. (1987a) The importance of the mechanisms of interspecific competition. *The American Naturalist*, **129**, 769–74.

Tilman, D. (1987b) On the meaning of competition and the mechanisms of competitive superiority. *Functional Ecology*, **1**, 304–15.

Tilman, D. (1988) *Plant Strategies and the Structure and Dynamics of Plant Communities*, Princeton University Press, Princeton, New Jersey.

Tilman, D., Kilham, S. S. and Kilham, P. (1984) A reply to Sell, Carney and Fahnenstiel. *Ecology*, **65**, 328–32.

Toffler, A (1984) Foreward: Science and Change. In *Order Out of Chaos*, (eds. I. Prigogine and I. Stengers), New Science Library, Shambala, Boulder, Colorado, xi–xxvi.

Trungpa, C. (1976) *The Myth of Freedom*, Shambala, Boston.

Turkington, R. and Mehrhoff, L. A. (1989) The role of competition in structuring pasture communities. In *Perspectives on Plant Competition* (eds J. Grace and D. Tilman), Academic Press, New York.

Underwood, A. J. (1978) The detection of non-random patterns of distribution of species along an environmental gradient. *Oecologia*, **36**, 317–26.

Underwood, A. J. and Denley, E. J. (1984) Paradigms, explanations and generalizations in models for the structure of intertidal communities on rocky shores. In *Ecological Communities: Conceptual Issues and the Evidence* (eds D. R. Strong, D. Simberloff, L. G. Abele and A. B. Thistle), Princeton University Press, Princeton, New Jersey, pp. 151–80.

Usher, M. B. (1973) *Biological Management and Conservation*, Chapman and Hall, London.

Vandermeer, J. H. (1970) The community matrix and the number of species in a community. *The American Naturalist*, **104**, 73–83.

Vandermeer, J. H. (1972) Niche theory. *Annual Review of Ecology and Systematics*, **3**, 107–32.

van der Valk, A. G. (1981) Succession in wetlands: a Gleasonian approach. *Ecology*, **62**, 688–96.

Vinson, S. B. (1976) Host selection by insect parasitoids. *Annual Review of Entomology*, **21**, 109–33.

Voltaire [Arouet, F.-M.] (1759) *Candide*, reprinted 1963 as Gay, P. *Voltaire's Candide: a Bilingual Edition*, St Martin's Press, New York.

Waage, J. K. (1979) Dual function of the damselfly penis: sperm removal and transfer. *Science*, **203**, 916–18.

Waage, J. K. (1984) Sperm competition and the evolution of odonate mating systems. In *Sperm Competition and the Evolution of Animal Mating Systems* (ed. R. L. Smith), Academic Press, Orlando, Florida.

Wallace, J. W. and Mansell, R. L. (eds) (1976) *Biochemical Interaction between Plants and Insects*, Recent Advances in Phytochemistry, Vol. 10, Plenum Press, New York.

Watkinson, A. R. (1985a) Plant responses to crowding. In *Studies on Plant Demography* (ed. J. White), Academic Press, London.

Watkinson, A. R. (1985b) On the abundance of plants along an environmental gradient. *Journal of Ecology*, **73**, 569–78.

Watts, A. H. (1958) *The Spirit of Zen*, Grove Press, New York.

Weaver, J. E. and Clements, F. E. (1929) *Plant Ecology*, McGraw-Hill, New York.

Weiner, J. (1986) How competition for light and nutrients affects size variability in *Ipomea tricolor* populations. *Ecology*, **67**, 1425–7.

Weiner, J. and Thomas, S. C. (1986) Size variability and competition in plant monocultures. *Oikos*, **47**, 221–2.

Weins, J. A. (1977) On competition and variable environments. *American Scientist*, **65**, 590–7.

Weins, J. A. (1981) Single sample surveys of communities: are the revealed patterns real? *The American Naturalist*, **117**, 90–8.

Weins, J. A. (1983) Avian community ecology: an iconoclastic view. In *Perspectives in Ornithology* (eds A. H. Brush and G. A. Clark), Cambridge University Press, Cambridge.

Weldon, C. W. and Slauson, W. L. (1986) The intensity of competition versus its importance: an overlooked distinction and some implications. *The Quarterly Review of Biology*, **61**, 23–44.

Werner, P. A. (1979) Competition and coexistence of similar species. In *Topics in Plant Population Biology* (eds O. T. Solbrig, S. Jain, G. G. Johnson and P. H. Raven), Columbia University Press, New York.

Westoby, M. (1984) The self-thinning rule. *Advances in Ecological Research*, **14**, 167–225.

White, P. S. (1979) Pattern, process and natural disturbance in vegetation. *The Botanical Review*, **45**, 229–99.

Whittaker, R. H. (1956) Vegetation of the Great Smoky Mountains. *Ecological Monographs*, **26**, 1–80.

Whittaker, R. H. (1967) Gradient analysis of vegetation. *Biological Reviews*, **42**, 207–64.

Whittaker, R. H. (1969) New concepts of kingdoms of organisms. *Science*, **163**, 150–60.

Whittaker, R. H. (1975) *Communities and Ecosystems*, 2nd edn, Macmillan, London.

Whittaker, R. H. and Levin, S. A. (1975) *Niche Theory and Application*, Dowden, Hutchinson and Ross, Stroudsburg.

Whittington, R. (1984) Laying down the $-3/2$ power law. *Nature*, **311**, 217.

Widden, P. (1984) The effects of temperature on competition for spruce needles among sympatric species of *Trichoderma*. *Mycologia*, **76**, 873–83.

Wilbur, H. M. (1972) Competition, predation and the structure of the *Ambystoma–Rana sylvatica* community. *Ecology*, **53**, 3–21.

Wilson, E. O. (1975) *Sociobiology*, Belknap Press, Harvard University Press, Cambridge, Massachusetts.

Wilson, E. O. (1978) *On Human Nature*, Harvard University Press, Cambridge, Massachusetts.

Wilson, J. B. (1988) Shoot competition and root competition. *Journal of Applied Ecology*, **25**, 279–96.

Wilson, S. D. (1986) Experimental evaluation of plant zonation patterns along an exposure gradient. PhD thesis, University of Ottawa.

Wilson, S. D. and Keddy, P. A. (1985) Plant zonation on a shoreline gradient: physiological response curves of component species. *Journal of Ecology*, **73**, 851–60.

Wilson, S. D. and Keddy, P. A. (1986a) Measuring diffuse competition along an environmental gradient: results from a shoreline plant community. *The American Naturalist*, **127**, 862–9.

Wilson, S. D. and Keddy, P. A. (1986b) Species competitive ability and position along a natural stress/disturbance gradient. *Ecology*, **67**, 1236–42.

Wimsatt, W. C. (1982) Reductionistic research strategies and their biases in the units of selection controversy. In *Conceptual Issues in Ecology* (ed. E. Saarinen), D. Reidel, Dordrecht, pp. 155–201.

Worster, D. (1977) *Nature's Economy*, Cambridge University Press, Cambridge.

Wright, S. J. and Biehl, C. C. (1982) Island biogeographic distributions: Testing for random, regular, and aggregated patterns of species occurrence. *The American Naturalist*, **119**, 345–57.

Yoda, K., Kira, T., Ogawa, H. and Hozumi, K. (1963) Self-thinning in overcrowded pure stands under cultivated and natural conditions. *Journal of Biology/Osaka City University*, **14**, 107–29.

Yodzis, P. (1978) *Competition for Space and the Structure of Ecological Communities*, Springer-Verlag, Berlin.

Yodzis, P. (1986) Competition, mortality and community structure. In *Community Ecology* (eds J. Diamond and T. J. Case), Harper and Row, New York, pp. 480–91.

Zukav, G. (1979) *The Dancing Wu Li Masters*, Morrow, New York.

Index